School Mathematics

Combined Volume for High Schools

Algebra, Geometry, Trigonometry, Calculus, and Real World Applications.

Chee Leong Ching

Sun Jie

Yink Loong Len

May Han Thong

Eksis Waiz

An Imprint of the Simplicity Research Institute, Singapore.
www.sribooks.simplicitysg.net

School Mathematics:
Combined Volume for High Schools.
SRI Books,
an imprint of the Simplicity Research Institute, Singapore.
www.sribooks.simplicitysg.net
email: enquiry@simplicitysg.net

SriBooks

Contents

Preface

This book consolidates in six chapters the Volumes 1 through 6 of the School Mathematics series published by SRI Books. It contains brief review notes, examples with detailed solutions, and test questions with answers. We hope the reader finds the material here useful.

Most of the questions have been selected, with some modification, from the books *Integrated Mathematics for Explorers by* Adeline Ng and R. Parwani, and *Real World Mathematics* by W.K. Ng and R. Parwani. The solutions have been edited from the corresponding *Solutions Manuals* by C.L. Ching and Sun Jie, and Y.L. Len and M.H. Thong. You can browse through those books online at **www.simplicitysg.net/books**.

You can also download relevant notes and a detailed formula list from the website indicated above.

Singapore,
May 2017.

Other titles by SRI Books and the Simplicity Research Institute

Integrated Mathematics for Explorers, by Adeline Ng and Rajesh R. Parwani
This book for mathematics explorers introduces topics through real-life examples, accompanied by exercises, problems, challenges, puzzles, and suggested investigations.

Solutions Manual: Integrated Mathematics for Explorers, by Chee Leong Ching and Sun Jie
This book provides solutions to selected questions from the previous text.

Real World Mathematics, by Wei Khim Ng and Rajesh R. Parwani
This resource is for those interested in real-world applications of mathematics. The problems are suitable for those in high schools and liberal arts colleges, with accompanying notes summarising the background for the questions. Later chapters discuss the scientific method and mathematical modelling.

Solutions Manual: Real World Mathematics, by Yink Loong Len and May Han Thong
This book provides solutions to selected questions from the previous text.

Simplicity in Complexity: An Introduction to Complex Systems, by Rajesh R. Parwani
Topics covered include: Self-organisation, emergence, agent-based simulations, complex networks, phase plane plots, fractals, chaos, and measures of complexity. Emphasis is placed on clarifying common misconceptions.

Physics: Volume One, Classical Foundations, by Rajesh R. Parwani
This book is for enthusiasts in school and beyond. Topics covered include: Physics as information compression, the principle of relativity, cause and effect, symmetries and conservation laws, order and disorder, the arrow of time, fields, waves, spontaneous symmetry breaking, and emergence.

View sample pages of all books at
www.simplicitysg.net/books.

Available from the publisher's webstore
www.store.simplicitysg.net,
and other outlets.

Contact: enquiry@simplicitysg.net.

Chapter 1

Elementary Mathematics

1.1 Review Notes

Consider the repeated multiplication $5 \times 5 \times 5$. A shorthand for that product is 5^3, a notation that is not only compact but also allows for rapid calculations using the rules that we summarise below.

More generally, if one writes $y = b^x$ for $b \neq 1$ and $b > 0$, then b is called the **base** while x is the **exponent** which need not be an integer. The function $y = b^x$ is sometimes referred to as an indicial function and the exponent x as the index or power.

Integral exponents occur in our familiar decimal notation which uses the base 10. For example, 352 is $3 \times 10^2 + 5 \times 10^1 + 2 \times 10^0$. That same number may be written in **scientific notation (standard form)** as 3.52×10^2.

More generally, any non-zero number may be written in the form $\pm A \times 10^p$ where $1 \leq A < 10$ and p is an integer.

Fractional exponents are useful too. For example $2^{1/2} = \sqrt{2}$.

Given a quadratic equation $ax^2 + bx + c = 0$, its roots may be found by completing the square: Multiply the equation by $4a$, add b^2 to both sides and rearrange to get $4a^2x^2 + 4abx + b^2 = b^2 - 4ac$, or $(2ax + b)^2 = \Delta$ where

$$\Delta \equiv b^2 - 4ac \tag{1.1}$$

is called the **discriminant**. The solutions are therefore given by

$$x = \frac{-b \pm \sqrt{\Delta}}{2a}. \tag{1.2}$$

The two roots are real if and only if $\Delta \geq 0$; the case $\Delta = 0$ corresponds to a repeated root.

A different way of completing the square is often useful: $ax^2 + bx + c = a\left(x^2 + \frac{b}{a}x + \frac{c}{a}\right) = a\left(x + \frac{b}{2a}\right)^2 + \left(c - \frac{b^2}{4a}\right)$. This form shows that the quadratic equation has its extremal value at $x = -b/(2a)$, the point being a maximum if $a < 0$ and a minimum if $a > 0$ (this is discussed more below).

What does the curve $y = ax^2 + bx + c$ look like? It is determined by the sign of a: When $a > 0$, we see that y is positive and increasing for very large values of $|x|$, so the curve must have a minimum point. Next, by completing the square, or equivalently by looking at the discriminant, we can determine if the curve has any intersection with the $y = 0$ axis (that is, any real roots). Similarly, the case $a < 0$ leads to a "hump" shaped curve which reaches a maximum value.

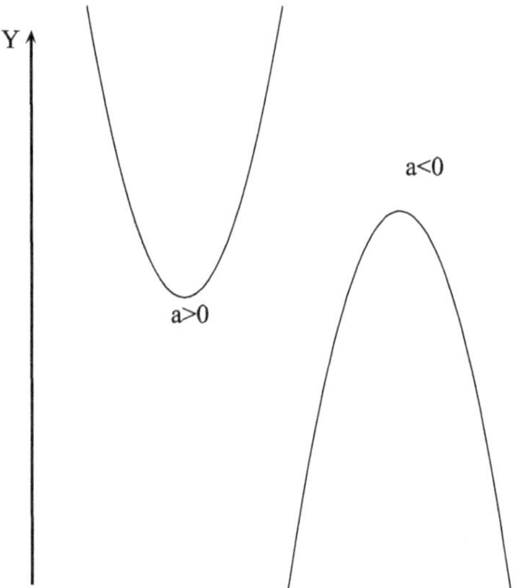

Figure 1.1: The shape of the parabola $y = ax^2 + bx + c$.

1.1.1 Relations and Properties

We list here some useful identities involving exponents.

For $a, b > 0$,

$$a^{-1} = \frac{1}{a}. \tag{1.3}$$

$$a^0 = 1. \tag{1.4}$$

$$a^{xy} = (a^x)^y. \tag{1.5}$$

$$a^x a^y = a^{x+y}. \tag{1.6}$$

$$(ab)^x = a^x b^x. \tag{1.7}$$

It is useful to note that for $a > 1$, $y = a^x$ is positive and an increasing function of x along the real line.

1.2 Worked Examples

1. Express the following numbers in scientific notation:

   (a) -0.0031.

   (b) 123.456.

   (c) $3.23 \times 12.56 \times 0.02$.

   (d) $-24.31 \times 10^5 \times 0.07 \times 10^{-6}$.

– –

Solutions:
Recall that any non-zero number may be written in scientific notation: $\pm A \times 10^p$ where $1 \leq A < 10$ and p is an integer.

We have,

(a) $-0.0031 = -3.1 \times 10^{-3}$.
(b) $123.456 = 1.23456 \times 10^2$.
(c) $3.23 \times 12.56 \times 0.02 = 8.11376 \times 10^{-1}$.
(d) $-24.31 \times 10^5 \times 0.07 \times 10^{-6} = -1.7017 \times 10^{-1}$.

---

2. The distances of the Moon, Sun and the star Proxima Centauri from Earth are respectively 3.8×10^5, 1.5×10^8 and 4.0×10^{13} kilometres (km). Taking the speed of light to be 3.0×10^5 km/s, calculate the time it takes light to reach Earth from each of those bodies. Express your answer in seconds for the Moon, in minutes for the Sun and in years for Proxima Centauri.

- -

Solutions:
Recall that speed (v) is defined as the ratio of distance travelled to the time taken for the journey, $v = d/t$. Therefore $t = d/v$.

For the Moon-Earth case, the time taken is

$$\begin{aligned} t_{\text{moon}} &= \frac{3.8 \times 10^5 \text{ km}}{3.0 \times 10^5 \text{ km/s}} \\ &= \frac{3.8}{3} \text{ s} \\ &= 1.3 \text{ s.} \end{aligned}$$

By the same method,

$$\begin{aligned} t_{\text{sun}} &= \frac{1.5 \times 10^8 \text{ km}}{3.0 \times 10^5 \text{ km/s}} \\ &= \frac{1.5}{3.0} \times 10^{8-5} \text{ s} \\ &= 0.5 \times 10^3 \text{ s} \\ &= 500 \text{ s.} \end{aligned}$$

Since 1 min = 60 s, therefore

$$\begin{aligned} t_{\text{sun}} &= \frac{500}{60} \text{ min} \\ &= 8.3 \text{ min.} \end{aligned}$$

Similarly,

$$
\begin{aligned}
t_{\text{centauri}} &= \frac{4.0 \times 10^{13} \text{ km}}{3.0 \times 10^{5} \text{ km/s}} \\
&= \frac{4}{3} \times 10^{13-8} \text{ s} \\
&= \frac{4}{3} \times 10^{8} \text{ s.}
\end{aligned}
$$

Now, $60 \times 60 = 3600$ seconds make one hour, 24 hours make a day and 365 days make a year. Therefore

$$
\begin{aligned}
t_{\text{centauri}} &= \frac{\frac{4}{3} \times 10^{8} \text{ s}}{365 \times 24 \times 60 \times 60 \text{s/yr}} \\
&= 4.2 \text{ yrs.}
\end{aligned}
$$

---

3. Re-arrange each of the following sequences in order, from the smallest to the largest, without using a calculator. Justify your answers.

   (a) 7^5, $5^{7/2}$, 7^7, 5^5.

   (b) $(2^6)^{\frac{1}{3}}$, $\left(2^{\frac{1}{2}}\right)^{-4}$, $2^2 \times 2^{-3}$, $(2^2 \times 3^4)^{-\frac{1}{2}}$.

\- -

Solutions:
Recall that a^x is an increasing function of x for $a > 1$.

(a) We have $7^7 > 7^5$, and similarly $5^5 > 5^{7/2}$. Furthermore, $5^5 < 7^5$ since $1 < \left(\frac{7}{5}\right)^5$. Therefore we can arrange the terms in increasing order as

$$5^{7/2} < 5^5 < 7^5 < 7^7.$$

(b) *Recall that $(a^x)^y = a^{xy}$ and $a^x \times a^y = a^{x+y}$.*
We first rewrite all the numbers in same base,

$$
\begin{aligned}
(2^6)^{1/3} &= 2^2 . \\
(2^{1/2})^{-4} &= 2^{-2} . \\
2^2 \times 2^{-3} &= 2^{2-3} = 2^{-1} . \\
(2^2 \times 3^4)^{1/2} &= (2^2)^{-1/2} \times (3^4)^{-1/2} \\
&= 2^{-1} \times 3^{-2}.
\end{aligned}
$$

We have $2^{-2} < 2^{-1} < 2^2$. Also, $2^{-1} \times 3^{-2} < 2^{-1} \times 2^{-1} = 2^{-2}$. Therefore,

$$(2^2 \times 3^4)^{-1/2} < (2^{1/2})^{-4} < 2^2 \times 2^{-3} < (2^6)^{1/3}.$$

---

4. Simplify the following expression as much as possible without using a calculator: $(2x^2)^{-\frac{1}{2}}\ (3x^{-5})^{-1}$.

- -

Solution:
Recall $(ab^x)^y = a^y b^{xy}$.
Therefore,

$$\begin{aligned} (2x^2)^{-1/2}\ (3x^{-5})^{-1} &= 2^{-1/2} \times x^{-1} \times 3^{-1} \times x^5 \\ &= \frac{1}{3\sqrt{2}}\ x^4. \end{aligned}$$

---

5. Solve each of the following equations for x without using a calculator:

   (a) $5^{2x-1} = 1/25$.
   (b) $3^{2-x}(2^2 \times 3^{2x+1}) = 4/9$.

- -

Solutions:
Recall $a^x = a^y \Rightarrow x = y$.

(a) We write all terms in the same base and compare exponents:

$$\begin{aligned} 5^{2x-1} &= \frac{1}{25} \\ &= \frac{1}{5^2} \\ &= 5^{-2} \\ \Rightarrow 2x - 1 &= -2 \\ \therefore x &= -\frac{1}{2}. \end{aligned}$$

(b) Similarly,

$$\begin{aligned} 3^{2-x}(2^2 \times 3^{2x+1}) &= \frac{4}{9} \\ &= 2^2 \times 3^{-2} \\ 3^{2-x} \times 3^{2x+1} &= 3^{-2} \\ 3^{(2-x)+(2x+1)} &= 3^{-2} \\ 3^{x+3} &= 3^{-2} \\ \Rightarrow x + 3 &= -2 \\ \therefore x &= -5. \end{aligned}$$

---

6. A bank pays 3% compound interest per month on its fixed deposits. If \$ 10,000 is deposited, what would be the accumulated amount three months later?

- -

Solution:
Note: Compounding at that rate means that at the end of each month the total amount would be 1.03 *times the amount at the start of that month. The total amount (principal plus accumulated interest), in dollars, at the end of n months is given by* $P(n) = (1+0.03)^n P_0$ *where* P_0 *is the initial amount deposited.*

For $n = 3$ we have

$$\begin{aligned} P &= (1+0.03)^3(10^4) \\ &= \left[1 + 3(0.03) + 3(0.03)^2 + (0.03)^3\right](10^4) \\ &= 10{,}927.27. \end{aligned}$$

Answer: \$ 10,927.27.

---

7. A certain bank pays r% simple interest for deposits which are kept for at least 3 months. If \$ 10,000 was deposited with this bank for three months, what value of r would give the same return as the bank of the previous question?

- -

Solution:
The total accumulated amount would now be $P_2(n) = (1 + r/100)P_0$ where $P_0 = 10^4$ dollars is the initial amount deposited. We need

$$\begin{aligned} \left(1 + \frac{r}{100}\right)(10^4) &= 10{,}927.27 \\ \left(1 + \frac{r}{100}\right) &= 1.092727 \\ \frac{r}{100} &= 0.092727 \\ r &\approx 9.27\ . \end{aligned}$$

That is, the simple interest rate would have to be about 9.27%.

---

8. Simplify the following expression:

$$\frac{(p^2+2)(p-3)+5p^2+10}{p^2+2p}$$

Solution:

Let us simplify the denominator first, writing it as $p(p+2)$. This suggests we try to find similar factors in the numerator. We notice $5p^2+10=5(p^2+2)$ which reveals the p^2+2 common to the first term of the numerator. This looks promising:

$$\begin{aligned}\frac{(p^2+2)(p-3)+5p^2+10}{p^2+2p} &= \frac{(p^2+2)(p-3)+5(p^2+2)}{p(p+2)} \\ &= \frac{(p^2+2)(p-3+5)}{p(p+2)} \\ &= \frac{(p^2+2)(p+2)}{p(p+2)} \\ &= \frac{(p^2+2)}{p}.\end{aligned}$$

---

9. A symmetrical spherical shell is created by removing a central sphere of radius R from the inside of a solid metal sphere of radius $2R$.

   (a) What is the volume of the remaining spherical shell?

   (b) The shell from part (a) is melted and the metal used to construct a cone of circular base with radius R. Determine the height of the cone in terms of R.

Solution:

Recall the volume of a sphere: $4\pi r^3/3$ where r is the radius.

(a) The volume of the shell is

$$\begin{aligned}\frac{4\pi}{3}\left((2R)^3-R^3\right) &= \frac{4\pi}{3}\left(8R^3-R^3\right) \\ &= \frac{28\pi}{3}R^3\,.\end{aligned}$$

(b) The volume of the cone is (area of base) × (height)/3. Letting the height be h, we have

$$\begin{aligned}\frac{h}{3}\pi R^2 &= \frac{28\pi}{3}R^3 \\ \Rightarrow h &= 28R\,.\end{aligned}$$

---

10. A line segment is divided into two parts of length a and b such that $\frac{a}{b} = \frac{a+b}{a} \equiv \phi$. Determine the numerical value of ϕ, known as the "Golden Ratio".

- -

Solution:

Note: $ax^2 + bx + c = 0 \Rightarrow x = \dfrac{-b \pm \sqrt{b^2 - 4ac}}{2a}$.

Let $\frac{a}{b} = x$, we can then rewrite the equation in the question as

$$\begin{aligned} \frac{a}{b} &= \frac{a+b}{a} \\ &= 1 + \frac{a}{b} \\ \Rightarrow x &= 1 + \frac{1}{x}. \end{aligned}$$

Next, we solve for x,

$$\begin{aligned} x - \left(1 + \frac{1}{x}\right) &= 0 \\ \frac{x^2 - x - 1}{x} &= 0 \\ \Rightarrow x^2 - x - 1 &= 0 \\ \therefore x &= \frac{-(-1) \pm \sqrt{(-1)^2 - 4(-1)}}{2} \\ &= \frac{1 \pm \sqrt{5}}{2}. \end{aligned}$$

Since the "Golden Ratio" is defined as the ratio of lengths (positive quantities), it has to be positive definite. Hence we only accept the positive solution, $\phi = x = \dfrac{1+\sqrt{5}}{2}$.

---

11. Neo, is on a platform that is $y_0 = 10$ m above ground. He throws a ball vertically upwards from that point. The subsequent displacement of the ball relative to the ground is given by $y(t) = y_0 + ut - \frac{gt^2}{2}$ where u m/s is the initial vertical velocity of the ball, g m/s^2 the acceleration due to gravity and t the time in seconds. If $u = 5$ and $g = 10$,

    (a) Factorise the expression for $y(t)$ for the given constants and use it to find the time taken for the ball to reach the ground.

(b) By writing $y(t)$ in the form $y = a(t+b)^2 + c$, determine the maximum height reached by the ball and the time taken for it to reach that height.

(c) For how long did the ball stay 5 m or more above ground?

- -

Solutions:

(a) With the given information, we have

$$y(t) = 10 + 5t - 5t^2,$$

which can be factorised into the form $y(t) = (t+1)(-5t+10)$.
When the ball reaches the ground again, $y(t) = 0$, so

$$\begin{aligned} 0 &= (t+1)(-5t+10) \\ \therefore t &= -1 \text{ s or } 2 \text{ s}. \end{aligned}$$

Since t represents the time, which is positive, the only acceptable answer is $t = 2$ s.

(b) Let us complete the square,

$$\begin{aligned} y(t) &= -5t^2 + 5t + 10 \\ &= -5(t^2 - t - 2) \\ &= -5\left[(t-1/2)^2 - 1/4 - 2\right] \\ &= -5\left[(t-1/2)^2 - 9/4\right] \\ &= -5(t-1/2)^2 + 45/4. \end{aligned} \tag{1.8}$$

We see that the first term on the right hand side of (1.8) is never positive, $-5(t-1/2)^2 \leq 0$. So, $y(t)$ will be maximum when that term vanishes.
The maximum height is $y_{\text{max}} = 45/4$ m and it is reached at $t = 1/2$ s.

(c) The ball starts at $t = 0$ from a platform which is 10 m above ground. It will move upwards but will eventually fall towards the ground. To find when it reaches $y = 5$ m, we solve

$$\begin{aligned} 5 &= -5t^2 + 5t + 10 \\ 0 &= -5t^2 + 5t + 10 - 5 \\ 0 &= -5t^2 + 5t + 5 \\ 0 &= -5(t^2 - t - 1) \\ \Rightarrow 0 &= t^2 - t - 1 \\ \therefore t &= \frac{-(-1) \pm \sqrt{1 - 4(-1)}}{2} \\ &= \frac{1 \pm \sqrt{5}}{2}. \end{aligned}$$

Since $t \geq 0$, we have to choose $t = (1 + \sqrt{5})/2$ s.

---

12. Solve the following equation for x: $\dfrac{1}{1-x}+\dfrac{1}{x+2}=3.$

- -

Solution:
We combine all terms over one common denominator,

$$\begin{aligned}
3 &= \frac{1}{1-x}+\frac{1}{x+2} \\
0 &= \frac{1}{1-x}+\frac{1}{x+2}-3 \\
0 &= \frac{(x+2)+(1-x)-3(1-x)(x+2)}{(1-x)(x+2)} \\
0 &= \frac{(x+2)+(1-x)-3(-x^2-x+2)}{(1-x)(x+2)} \\
0 &= \frac{3x^2+3x-3}{(1-x)(x+2)}\ .
\end{aligned}$$

Since the numerator must be zero,

$$\begin{aligned}
\Rightarrow x^2+x-1 &= 0 \\
\therefore\ x &= \frac{-1\pm\sqrt{1-4(-1)}}{2} \\
&= (-1\pm\sqrt{5})/2\ .
\end{aligned}$$

---

13. Find the range of values of x for which:

    (a) $2 \geq 5x + x^2$.

    (b) $4x \geq (x-1)(x-3)$.

- -

Solutions:

(a) We begin by bringing all the terms to one side and factorising the quadratic expression using its roots.

$$\begin{aligned}
2 &\geq 5x+x^2 \\
0 &\geq x^2+5x-2.
\end{aligned}$$

Let $f_1(x) = x^2 + 5x - 2$. Its roots are

$$x = \frac{-5 \pm \sqrt{25 - 4(-2)}}{2} = \frac{-5 \pm \sqrt{33}}{2}.$$

Let $x_2 = (-5 + \sqrt{33})/2$ and $x_1 = (-5 - \sqrt{33})/2$; see Fig. 1.2. Then

$$\begin{aligned} & 0 \geq (x - x_1)(x - x_2) \\ & \Rightarrow x_1 \leq x \leq x_2 \\ \therefore \quad & \frac{-5 - \sqrt{33}}{2} \leq x \leq \frac{\sqrt{33} - 5}{2}. \end{aligned}$$

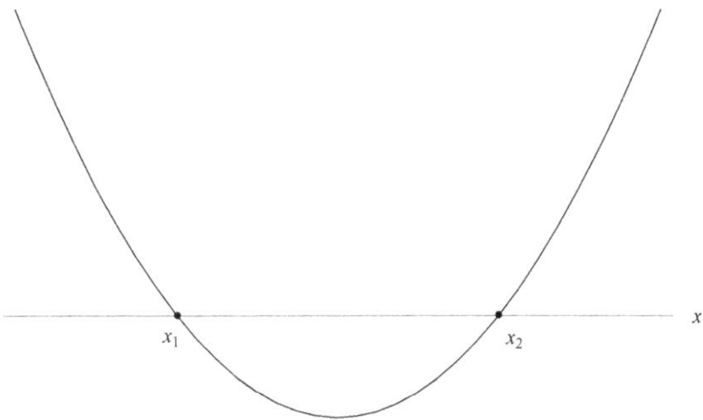

Figure 1.2: Parabola intersecting the x-axis.

(b) We proceed as in the previous example.

$$\begin{aligned} 4x & \geq (x - 1)(x - 3) \\ 0 & \geq (x - 1)(x - 3) - 4x \\ & \geq x^2 - 8x + 3 \end{aligned} \tag{1.9}$$

Let $f_2(x) = x^2 - 8x + 3$. The roots of $f_2(x)$ are

$$\begin{aligned} x & = \frac{8 \pm \sqrt{64 - 4(3)}}{2} \\ & = 4 \pm \sqrt{13}. \end{aligned}$$

Let $x_2 = 4 + \sqrt{13}$ and $x_1 = 4 - \sqrt{13}$. Thus,

$$\begin{aligned} & 0 \geq (x - x_1)(x - x_2) \\ & \Rightarrow x_1 \leq x \leq x_2 \\ \therefore \quad & 4 - \sqrt{13} \leq x \leq 4 + \sqrt{13}. \end{aligned}$$

14. If $f(x) = x^2 + bx + c$, determine the constants b, c for the separate exercises below and sketch the curve $y = f(x)$:

    (a) $f(x)$ has a double root at $x = 5$.

    (b) $f(x)$ has roots at $x = -1$ and $x = 2$.

- -

Solutions:

(a) If $f(x)$ has a double root at $x = 5$, it implies that $f(x) = a(x-5)^2 = a(x^2 - 10x + 25)$, with a a constant. Comparing with the given expression, $f(x) = x^2 + bx + c$, we deduce that $b = -10,\ c = 25$.

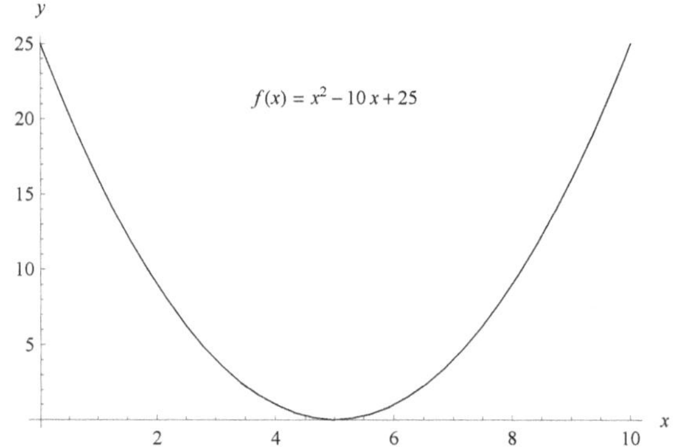

Figure 1.3: Plot of $f(x) = x^2 - 10x + 25$. There is a double root at $x = 5$.

(b) If $f(x)$ has roots at $x = -1$ and $x = 2$, it must be of the form $f(x) = a(x+1)(x-2) = a(x^2 - x - 2)$, with a a constant. Comparing with the given expression gives $b = -1,\ c = -2$.

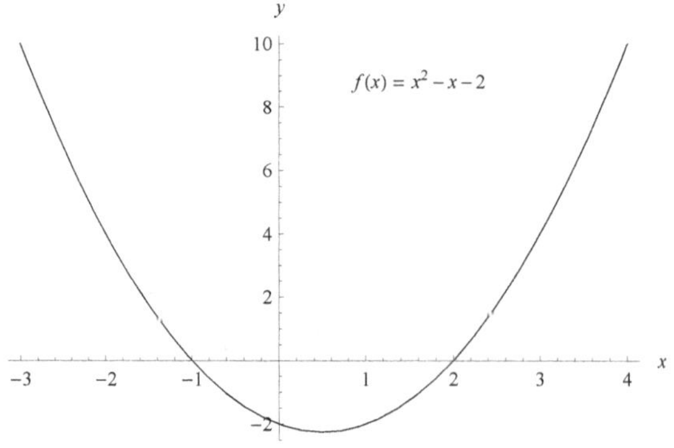

Figure 1.4: Plot of $f(x) = x^2 - x - 2$. The roots are at $x = -1$ and $x = 2$.

---

15. For each $f(x)$ determined in the previous question, sketch $y = |f(x)|$ and determine the range that y takes for x between 0 and the positive root.

Solutions:

(a) Since $f(x) = x^2 - 10x + 25 = (x-5)^2$ is always positive definite, the modulus graph is exactly same as the original graph, $|f(x)| = f(x)$, (see Fig.1.3). From the plot, the range that y takes for x between 0 and the positive root $x = 5$ is $0 \le y \le 25$.

(b) The plot of $|f(x)|$, where $f(x) = x^2 - x - 2$, is shown below.

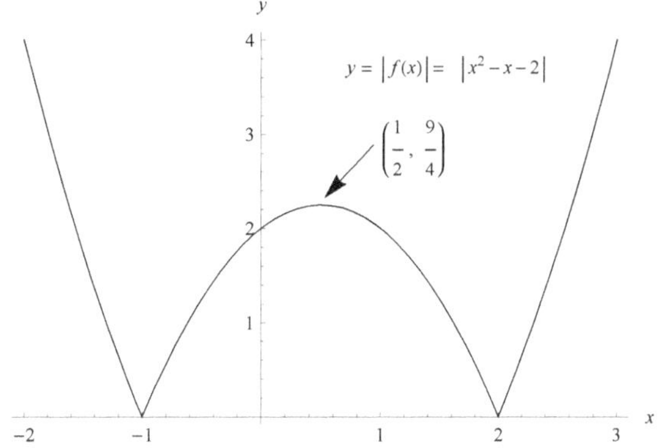

Figure 1.5: Plot of $|f(x)|$ where $f(x) = x^2 - x - 2$.

The local maximum of the graph $|f(x)|$ can be determined as follows,

$$\begin{aligned} f(x) &= |x^2 - x - 2| = |(x-1/2)^2 - 1/4 - 2| \\ &= |(x-1/2)^2 - 9/4|. \end{aligned}$$

Thus the maximum occurs at $x = 1/2$ and $y_{\max} = 9/4$. The range that y takes for x between 0 and the positive root $x = 2$ is $0 \le y \le 9/4$.

---

16. If $f(x) = x^2 + bx + 2b$, determine constraints on the constant b for the separate exercises below:

(a) $f(x)$ has no real roots.

(b) $f(x)$ has a root at $x = -1$.

\- -

Solutions:
Recall that a quadratic equation $Ax^2 + Bx + C = 0$ *has discriminant* $\Delta = B^2 - 4AC$. *The discriminant of* $f(x) = x^2 + bx + 2b$ *is* $b^2 - 8b = b(b - 8)$.

(a) For $f(x)$ to have no real roots, the discriminant must be negative,

$$\begin{aligned} & 0 > b(b-8) \\ \therefore \quad & 0 < b < 8. \end{aligned}$$

(b) If $f(x)$ has a root at $x = -1$, then $f(-1) = 0$, implying

$$\begin{aligned} 0 &= 1^2 + b(-1) + 2b \\ \Rightarrow b + 1 &= 0 \\ \therefore b &= -1. \end{aligned}$$

1.3 Test Yourself

1. A solid sphere occupies a volume of exactly 5000 cm^3. Determine its surface area to three significant figures and express your answer in standard form.

2. In Worked Example (11) of Chap. 2, if Neo were to throw the ball the same way on the Moon, where the g value is 1/6-th of that on Earth, what would the answers to the various parts of the question be?

3. Re-arrange the following three numbers in order, from the smallest to the largest, without using a calculator: $8^{1/3}$, $4^{-\frac{3}{2}}$, $\left(\frac{1}{2}\right)^{-2}$. Justify your answer.

4. A bank pays 4% compound interest per month on its fixed deposits. If a man deposits \$ X, and accumulates \$ 21,500 at the end of 6 months, what is X? Express your answer to the nearest dollar.

5. Rewrite the expression below to obtain an explicit form for x in terms of the constants a and b. Factorise your expression where possible.

$$\frac{1}{x} = \frac{1}{a^2} - \frac{1}{b^2} .$$

6. Find the values of x for which $2x + 5 \leq (3 - x)^2$.

7. If $f(x) = x^2 + bx + 2b$, determine constraints on the constant b if $f(x)$ has two distinct real roots.

8. Challenge: If the sum of the first n even natural numbers is A, and the sum of the first n odd natural numbers is B, show that $A - B = n$. (Note: A natural number is a positive integer).

1.4 Answers to Test

1. 1.41×10^3 cm^2.

2. (a) $t = (3 + \sqrt{21})$ s.

   (b) $y_{\mathrm{max}} = 35/2$ m when $t = 3$ s.

   (c) $t = (3 + \sqrt{15})$ s.

3. $4^{-3/2} < 8^{1/3} < \left(\frac{1}{2}\right)^{-2}$.

4. \$ 16,992.

5. $\dfrac{(ab)^2}{(b-a)(b+a)}$.

6. $x \geq 2(2+\sqrt{3})$ or $x \leq 2(2-\sqrt{3})$.

7. $b > 8$ or $b < 0$.

8. You can pair consecutive integers in the difference, for example, for $n = 3$, $(2+4+6) - (1+3+5) = (2-1) + (4-3) + (6-5) = 1 + 1 + 1 = 3$, and generalise. Or, use the formula for the sum of an arithmetic progression.

Did You Know?

The sum of the first n odd integers is n^2 .

For example, $1 + 3 + 5 = 3^2$.

Can you prove the general result below?

$$\sum_{k=1}^{n} (2k-1) = n^2 \ .$$

Chapter 2

Basic Algebra, Coordinate Geometry and Trigonometry

2.1 Review: Simultaneous equations

In simple cases, two simultaneous equations may be reduced to one equation by eliminating one common variable between them. However, even when this is possible, the resulting equation might be too difficult to solve analytically, as in the example of the two equations $y = 2^{-x}$ and $y - x - 1/2 = 0$. For such cases one may find an approximate solution using **graphical methods**: Plot the two curves $y = 2^{-x}$ and $y = x + 1/2$ on the same graph and see where they intersect.

What about inequalities? If there is a combination of equations and inequalities, then typically the boundaries of the solution set are first determined by treating all of them as equalities. For example $y = x^2$ and $y < x - 1$ is equivalent to $x - 1 > x^2$ or $x^2 - x + 1 < 0$. Solving the equation $x^2 - x + 1 = 0$ and making a sketch will help determine the solution set for the original problem involving the inequality.

2.2 Review: Trigonometry

Given a right-angled triangle labelled by its vertices (and thus angles) ABC, with $C = 90°$, the trigonometric functions $\sin A$ ("**sine A**") and $\cos A$ ("**cosine A**") are initially defined by the ratios $\sin A = a/c$ and $\cos A = b/c$, where the small case letters denote lengths opposite the corresponding angles, see Fig.(2.1).

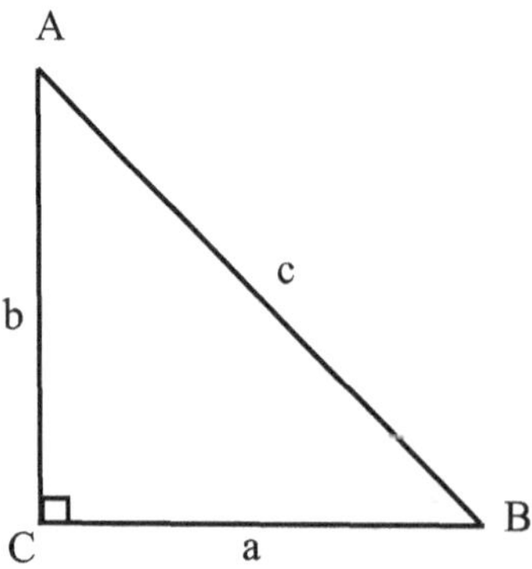

Figure 2.1: A right-angled triangle.

From the above triangle one deduces the identity $\sin(90° - A) = \cos A$ and by using Pythagoras theorem you can verify another identity: $\sin^2 A + \cos^2 A = 1$.

For the generalisation to angles larger than 90°, consider a circle with centre at O and radius R as shown in Fig.(2.2). Let P be a point on the circumference with coordinates (x, y) and denote by θ the angle that the line OP makes with the positive $x-$axis.

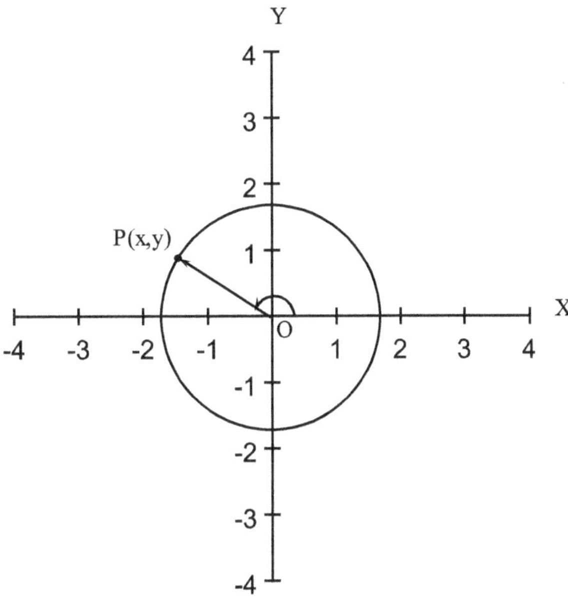

Figure 2.2: A circle is used to define the trigonometric functions for all angles.

By **convention**, angles are positive when one moves counter-clockwise away from the positive x-axis, and vice versa. Define $\sin\theta = y/R$ and $\cos\theta = x/R$. Note that $\sin\theta$ is positive in the first quadrant $(0 \leq \theta \leq \pi/2)$ and second quadrant $(\pi/2 \leq \theta \leq \pi)$ while $\cos\theta$ is positive in the first and fourth quadrant $(3\pi/2 \leq \theta \leq 2\pi)$. So, $\sin(-\theta) = -\sin(\theta)$, $\cos(-\theta) = \cos(\theta)$ and both functions range over the interval $[-1, 1]$.

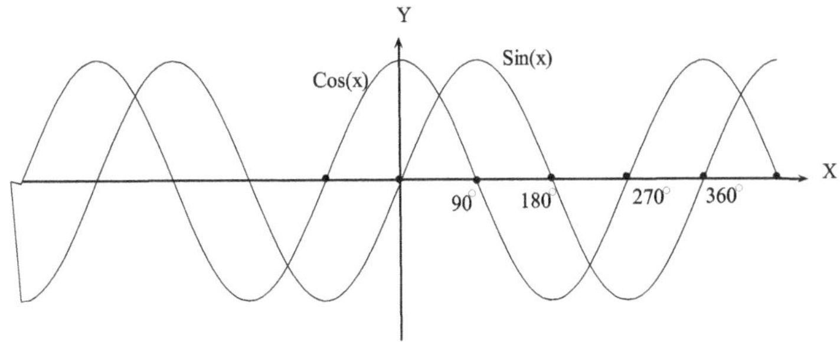

Figure 2.3: The sin and cosine functions

With the above definition of the trigonometric functions for all angles, we see that they are **periodic**, with period 360°, that is $\sin(\theta + 360°) = \sin(\theta)$ and $\cos(\theta + 360°) = \cos(\theta)$.

It is this periodic nature of trigonometric functions that makes them very useful in the description of numerous oscillatory phenomena in Nature, science and engineering, such as the motion of tides, waves, and alternating currents. Phenomena that are described by an equation of the form $y(x) = A + B\sin(kx + C)$ are called **sinusoidal**.

The other common trigonometric function, $\tan A$ ("**tangent A**") is defined as the ratio $\frac{\sin A}{\cos A}$; it has a period of 180° and ranges over the whole real line.

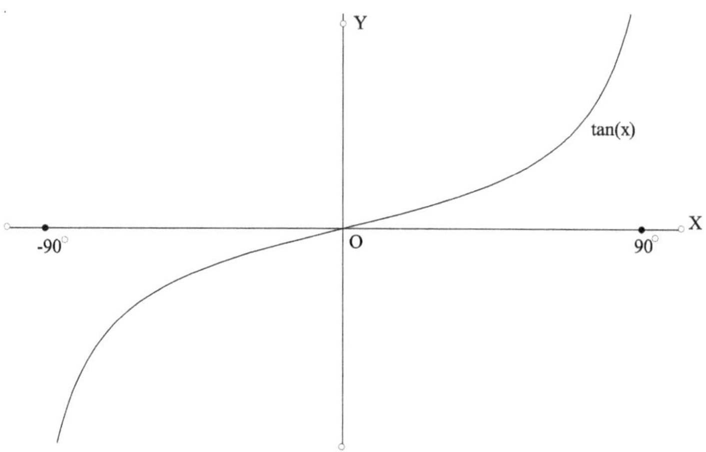

Figure 2.4: The "tan" function

The notation for the **inverse** sine function is $y = \sin^{-1} x$ which refers to values of y for which $\sin y = x$. Since the trigonometric functions are periodic, the inverse functions are multi-valued and so usually one restricts the inverse functions to their **principal values** (so that there is a unique inverse): For $y = \cos^{-1} x$ the values lie in the range $0 \leq y \leq \pi$ while for $\sin^{-1} x$ and $\tan^{-1} x$ functions the range is $-\pi/2 \leq y \leq \pi/2$.

Finally, we note that for angle measurements, the dimensionless unit **radian** is useful. Recall that in radians, $\theta = s/R$ where s is the arc length of circle subtended by that angle. Therefore 2π radians equals 360°.

2.2.1 Properties of Triangles

Bear in mind that the relations in the previous sub-section are properties of the trigonometric functions, holding for all angles.

One may also derive trigonometric relations which hold specifically for triangles. For a right-angled triangle, which is quite special, a popular mnemonic is "**soh-cah-toa**", which summarises the rules $\sin = \frac{Opposite\ side}{Hypotenuse}$, $\cos = \frac{Adjacent\ side}{Hypotenuse}$ and $\tan = \frac{Opposite\ side}{Adjacent\ side}$.

Other trigonometric relations for triangles usually involve both angles and lengths. Consider the triangle ABC in Fig.(2.5).

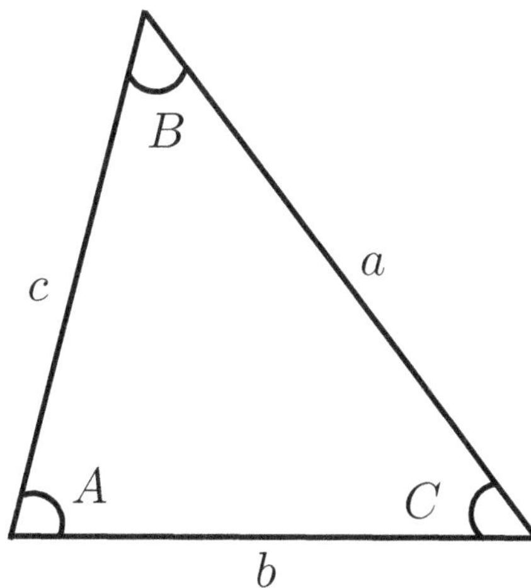

Figure 2.5: Labelling of generic triangle.

Its area A is given by

$$A = \frac{1}{2}ab\ \sin C = \frac{1}{2}ac\ \sin B = \frac{1}{2}bc\ \sin A\,. \tag{2.1}$$

The **sine rule** for triangles is

$$\frac{c}{\sin C} = \frac{b}{\sin B} = \frac{a}{\sin A}\,. \tag{2.2}$$

The **cosine rule** is

$$c^2 = a^2 + b^2 - 2ab\ \cos C\,. \tag{2.3}$$

Of course you can permute the labels of the vertices and find analogous expressions involving $\cos A$ or $\cos B$.

2.2.2 Elevation, Depression and Bearing

Trigonometry is useful in surveying and navigation. We introduce some common terminology here.

If a point P is above the horizontal through the observation point O, then the angle that OP makes with the horizontal is the **angle of elevation** of P. Similarly if Q were below the horizontal then the corresponding angle would be termed **angle of depression**.

A **bearing** denotes a direction relative to North. It is usually expressed by a clockwise angle measured in degrees. For example, 065° would refer to the direction 65° clockwise from North. Such bearings are called "absolute bearings".

It is also convenient to use "relative bearings", which define an angle relative to a chosen axis, for example the axis along which an aircraft is pointing.

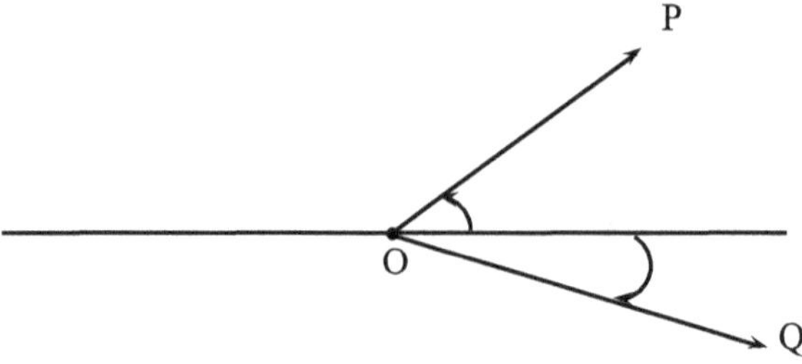

Figure 2.6: Elevation and depression angles relative to the horizontal.

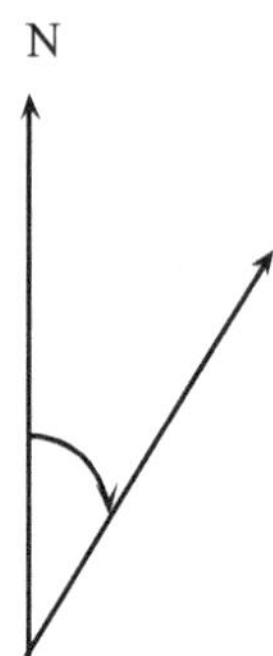

Figure 2.7: Absolute bearings are measured clockwise relative to North.

2.3 Review: Coordinate Geometry

A straight line has constant slope (gradient), so if (x, y) and (x_1, y_1) are any two points on the line, we may write

$$\frac{y - y_1}{x - x_1} = m = \tan\alpha\,, \tag{2.4}$$

where $0 \leq \alpha < \pi$ is the angle the line makes with the positive x-axis. The equation for a straight line may be written in alternate forms, for example as $ax + by + d = 0$, for some constants a, b, d or as

$$y = mx + c\ , \tag{2.5}$$

where m is the slope and c the intercept on the y-axis.

The straight line distance between point $(x_1,\ y_1)$ and $(x_2,\ y_2)$ is given by

$$\sqrt{(x_1 - y_1)^2 + (x_2 - y_2)^2}. \tag{2.6}$$

2.4 Worked Examples

1. Two boats of tourists are next to each other. The passengers of A say to the passengers of B: " If one of you comes over to our boat, then we will have twice as many passengers as you". However, the passengers of B reply with a better proposal to A: "If one of you comes over to our boat then each boat will have the same number of passengers. How many passengers are in each boat?

- -

Solution:
Denote the number of passengers in boats A and B by a and b respectively. From the question we have the following information

$$\begin{aligned} a+1 &= 2(b-1)\ , \\ a-1 &= b+1. \end{aligned}$$

Taking the difference between the two equations, we get

$$\begin{aligned} (a+1)-(a-1) &= 2(b-1)-(b+1) \\ 2 &= b-3 \\ \Rightarrow b &= 5. \end{aligned}$$

After substitution, $a = 2(b-1) - 1 = 2(5-1) - 1 = 7$.

---

2. The 200 students of a school are teamed up into groups of either 3 or 5. If the total number of teams formed is less than 50, determine a constraint on the number of teams with 3 members.

- -

Solution:
Denote the number of teams with three students and the number with five students by x and y respectively. Since we have in total 200 students, $3x + 5y = 200$, or $y = (200 - 3x)/5$. Also, we are told that total number of teams formed is less than 50, that is $x + y < 50$. By substituting y into the inequality we get

$$\begin{aligned} 50 &> x + \left(\frac{200-3x}{5}\right) \\ 50 &> \frac{5x+200-3x}{5} \\ 250 &> 5x + (200-3x) \\ 250-200 &> 5x-3x \\ 50 &> 2x \\ \therefore\ 25 &> x. \end{aligned}$$

So, the number of teams with three members must be less than 25.

---

3. Find the range of values of x which satisfy the following:

(a) $-3x + 3 > 2 - x$.

(b) $x - 1 > 3x - 4 > -2 - x$.

- -

Solutions:

(a) Given the inequality $-3x + 3 > 2 - x$, we can rearrange to obtain

$$\begin{aligned} -3x + 3 &> 2 - x \\ 3 - 2 &> -x + 3x \\ 1 &> 2x \\ \therefore \frac{1}{2} &> x. \end{aligned}$$

(b) We are given $x - 1 > 3x - 4 > -2 - x$. We can treat this as two simultaneous inequalities

$$\begin{aligned} x - 1 &> 3x - 4 \\ \text{and} \quad 3x - 4 &> -2 - x. \end{aligned}$$

We can re-arrange the terms to obtain two simplified inequalities,

$$\begin{aligned} 2x < 3 &\Rightarrow x < 3/2 \\ \text{and} \quad 4x > 2 &\Rightarrow x > 1/2. \end{aligned}$$

The common domain of these inequalities is $\frac{1}{2} < x < \frac{3}{2}$.

---

4. Solve the following simultaneous equations for x and y : $2y - 5x = 1$ and $y = 3x^2 - 2$.

- -

Solution:
We can substitute the second equation into the first one, and solve the resulting quadratic equation,

$$\begin{aligned} 1 &= 2(3x^2 - 2) - 5x \\ &= 6x^2 - 5x - 4 \\ \Rightarrow 0 &= 6x^2 - 5x - 5 \\ \therefore x &= \frac{5 \pm \sqrt{25 - 4(6)(-5)}}{12} \\ &= \frac{5 \pm \sqrt{145}}{12}. \end{aligned}$$

Then we get y by substitution, $y = (5x+1)/2$,

$$\begin{aligned} y &= \frac{\left(5 \times \frac{5\pm\sqrt{145}}{12} + 1\right)}{2} \\ &= \frac{5(5\pm\sqrt{145})+12}{24} \\ &= \frac{25 \pm 5\sqrt{145}+12}{24} \\ &= \frac{37 \pm 5\sqrt{145}}{24}. \end{aligned}$$

---

5. Find the values of x which satisfy $2x^2 - 3x - 4 < 0$ and $2 - x < 0$.

\- -

Solution:

The roots of $y = 2x^2 - 3x - 4 = 0$ are $x_\pm = \dfrac{3 \pm \sqrt{41}}{4}$. The domain of x that makes $y < 0$ lies between the roots (see Volume 1); that is

$$\frac{3-\sqrt{41}}{4} < x < \frac{3+\sqrt{41}}{4}.$$

Meanwhile, $2 - x < 0$ implies $x > 2$. Combining both conditions, we get $2 < x < \dfrac{3+\sqrt{41}}{4}$. (Note: Since $\sqrt{41} > 6$, $\dfrac{3-\sqrt{41}}{4} < 0 < 2$).

---

6. Students at a fund-raising fair sell three home-made drinks named, "Cool", "Wow" and "Zing" using the same natural ingredients but in different proportions as listed in the table below.

   (a) Construct a 3×3 matrix to represent that information.

   (b) All the ingredients have to be bought from one of two suppliers (A, B) whose prices (per glass of drink) are listed in the second table. Calculate, using appropriate matrix multiplication, how much each glass of drink would cost to make if it was made using materials wholly provided by one or the other supplier (assume water is provided free by the school and no labour or other costs are involved.)

   (c) The drinks made using ingredients from supplier A are sold at one stall, the "Apex", while the drinks made using ingredients from supplier B are sold at another stall, "Bravo". Each drink is sold at double its cost price and the number of glasses sold at each stall is listed in the table below. Calculate, with explanation, the total profit from the sales at (i) stall Apex , (ii) stall Bravo.

Ingredients table (in dimensionless units):

	Cool	Wow	Zing
Colouring	1	2	3
Flavouring	2	3	1
Sweetener	3	1	2

Ingredients cost (in dollars) per unit ingredient:

	Colouring	Flavouring	Sweetener
Supplier A	0.1	0.2	0.3
Supplier B	0.15	0.15	0.25

Sales table (number of glasses):

	Cool	Wow	Zing
Apex	5	10	20
Bravo	10	15	15

- -

Solutions:

(a) The matrix is

$$X = \begin{pmatrix} 1 & 2 & 3 \\ 2 & 3 & 1 \\ 3 & 1 & 2 \end{pmatrix}.$$

The columns represent the drinks while the rows represent the ingredients.

(b) The cost matrix is

$$Y = \begin{pmatrix} 0.1 & 0.2 & 0.3 \\ 0.15 & 0.15 & 0.25 \end{pmatrix}.$$

To obtain the cost of drinks made by using materials wholly from supplier A or B, we multiply Y and X,

$$\begin{aligned} YX &= \begin{pmatrix} 0.1 & 0.2 & 0.3 \\ 0.15 & 0.15 & 0.25 \end{pmatrix} \cdot \begin{pmatrix} 1 & 2 & 3 \\ 2 & 3 & 1 \\ 3 & 1 & 2 \end{pmatrix} \\ &= \begin{pmatrix} 1.4 & 1.1 & 1.1 \\ 1.2 & 1.0 & 1.1 \end{pmatrix}. \end{aligned}$$

Therefore the cost table is

	Cool	Wow	Zing
Supplier A	1.4	1.1	1.1
Supplier B	1.2	1.0	1.1

(c) Since the sale price is double the cost price, the profit matrix at each store for each drink is $P = \begin{pmatrix} 1.4 & 1.1 & 1.1 \\ 1.2 & 1.0 & 1.1 \end{pmatrix}$. At Apex, the total profit equals the first row of the profit matrix multiplied by the number of glasses sold,

$$\begin{pmatrix} 1.4 & 1.1 & 1.1 \end{pmatrix} \cdot \begin{pmatrix} 5 \\ 10 \\ 20 \end{pmatrix} = 40 \text{ dollars.}$$

Similarly, at Bravo the total profit is

$$\begin{pmatrix} 1.2 & 1.0 & 1.1 \end{pmatrix} \cdot \begin{pmatrix} 10 \\ 15 \\ 15 \end{pmatrix} = 43.5 \text{ dollars.}$$

---

7. The line $2x + 3y - 1 = 0$ intercepts the x-axis at P and the y-axis at Q. Determine

(a) The coordinates of P and Q.

(b) The length of PQ.

- -

Solutions:

(a) At P we have $y = 0$. Substitution into the equation of the line gives

$$\begin{aligned} 2x - 1 &= 0 \\ \Rightarrow x &= 1/2 \\ \therefore\ P &= (1/2, 0). \end{aligned}$$

At Q, $x = 0$, so

$$\begin{aligned} 3y - 1 &= 0 \\ \Rightarrow y &= 1/3 \\ \therefore\ Q &= (0, 1/3). \end{aligned}$$

(b) Using Eq.(2.6), the length is $\sqrt{(1/2 - 0)^2 + (0 - 1/3)^2} = \sqrt{13}/6$.

---

8. Four points on the plane are $A(1, a)$, $B(b, 2)$, $C(3, 5)$ and $D(4, 7)$. Determine

(a) The slope of the line passing through C and D, and its equation.

(b) The constants a and b, if the four points lie on a straight line.

– –

Solutions:

(a) From Eq.(2.4), the slope is $\dfrac{7-5}{4-3} = 2$. Let (x, y) be a point on the line, then using the point $(3, 5)$ as reference, we have

$$\begin{aligned} \frac{y-5}{x-3} &= 2 \\ y-5 &= 2x-6 \\ y &= 2x-1 \, . \end{aligned}$$

(b) The straight line is $y = 2x-1$. All points on the line, such as A and B, must satisfy that equation. For A, $a = 2(1)-1 = 1$, while for B, we have $2 = 2(b)-1 \Rightarrow b = 3/2$.

---

9. Neo decides to measure the height of building. He walks 30 m away from the base of the building, lies down on the ground and uses a home-made device to measure the angle 36° between the ground and the top of the building.

(a) What is the height of the building?

(b) If, lying at the same place, Neo used his device to look at a point P half-way up the building, what angle would he measure?

(c) If, lying at the same place, Neo used his device to look at a point Q on the building making an angle 18° with the ground, how high is Q above ground?

(d) Why are the locations of P and Q different even though 18° is half of 36°?

– –

Solution:

(a) See the figure below.

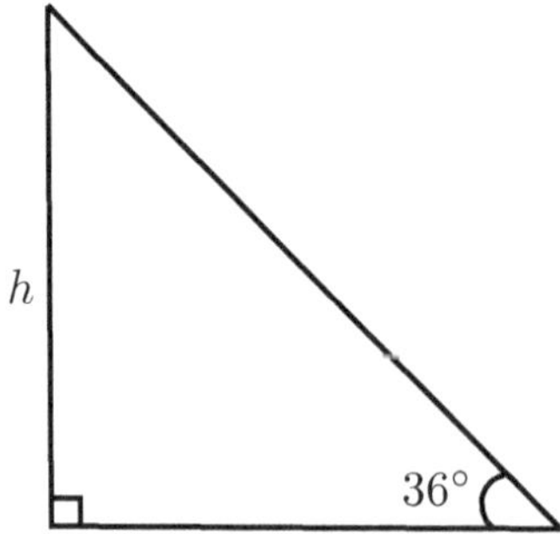

Figure 2.8: Figure for Question 9.

To find the height of the building, we take the tangent of the indicated angle,

$$\begin{aligned} \tan 36^0 &= \frac{h}{30} \\ \Rightarrow h &= 30\tan 36^0 = 21.80 \text{ m}. \end{aligned}$$

(b) Let the new angle subtended to point P be α,

$$\begin{aligned} \tan\alpha &= \frac{h/2}{30} \\ &= \frac{21.8}{2(30)} = 0.363 \\ \Rightarrow \alpha &= \tan^{-1} 0.363 = 19.96^0. \end{aligned}$$

(c) Let point Q be at a height h' above ground. Then

$$\begin{aligned} \tan 18^0 &= \frac{h'}{30} \\ \Rightarrow h' &= 30\tan 18^0 = 9.75 \text{ m}. \end{aligned}$$

(d) The locations of P and Q are different since the tangent function is not linear, for example, $\tan 2x \neq 2\tan x$. The graph (2.4) illustrates the non-linear behaviour of the tangent function.

---

10. The time-varying coordinates of a particle moving in a constant gravitational field (projectile motion) are given by $x(t) = Ut\cos\theta$ and $y(t) = Ut\sin\theta - \frac{gt^2}{2}$ (ignoring air resistance), where U is the initial velocity of the particle, g the acceleration due to gravity, and θ the launch angle relative to the horizontal..

   (a) Eliminate the time variable to obtain an equation for the trajectory, $y = y(x)$.

   (b) The range R is the horizontal distance between the start of the projectile and the place where it hits the ground. Find an expression for R.

   (c) Determine the maximum value of R as θ varies, keeping the other parameters fixed. What value of θ maximises R?

\- -

Solutions:

(a) We can use the equation of horizontal motion $x(t) = Ut\cos\theta$ to express time t as $t = \dfrac{x}{U\cos\theta}$ and substitute it into $y(t)$,

$$\begin{aligned} y(x) &= U\left(\frac{x}{U\cos\theta}\right)\sin\theta - \frac{g}{2}\left(\frac{x}{U\cos\theta}\right)^2 \\ &= x\tan\theta - \frac{gx^2}{2U^2\cos^2\theta} \\ &= x\tan\theta - \frac{gx^2\sec^2\theta}{2U^2} \end{aligned}$$

where $\sec\theta \equiv 1/\cos\theta$.

(b) When the particle hits the ground again, $y(x) = 0$. So we have

$$\begin{aligned} 0 &= y\big|_{x=R} \\ 0 &= R\tan\theta - \frac{gR^2\sec^2\theta}{2U^2} \\ &= R\left[\tan\theta - \frac{gR\sec^2\theta}{2U^2}\right] \\ \Rightarrow 0 = R &\quad \text{or} \quad 0 = \tan\theta - \frac{gR\sec^2\theta}{2U^2}. \end{aligned}$$

Since the range $R \neq 0$, we have to choose the second solution given by

$$\begin{aligned} 0 &= \tan\theta - \frac{gR\sec^2\theta}{2U^2} \\ \tan\theta &= \frac{gR\sec^2\theta}{2U^2} \\ \Rightarrow R &= \frac{2U^2\tan\theta}{g\sec^2\theta} \\ &= \frac{2U^2\frac{\sin\theta}{\cos\theta}}{g} \times \cos^2\theta \\ &= \frac{2U^2\sin\theta\cos\theta}{g} \\ &= \frac{U^2\sin 2\theta}{g} \end{aligned}$$

where we used the identity $\sin 2\theta \equiv 2\sin\theta\cos\theta$.

(c) From $R = \dfrac{U^2\sin 2\theta}{g}$ we see that the range is maximum when $\sin 2\theta$ is maximum (when the other parameters are kept fixed). Since the maximum value of the sine function is 1, this happens when $2\theta = \dfrac{\pi}{2} \Rightarrow \theta = \dfrac{\pi}{4}$. Hence the maximum range is $R_{\text{max}} = U^2/g$ and it is achieved when $\theta = \dfrac{\pi}{4}$.

---

11. Given $\sin x = 1/5$, and $0 < x < \pi/2$, find the exact values of $\cos x$ and $\tan x$.

– –

Solutions:
From the identity $\sin^2 A + \cos^2 A = 1$ we get

$$\begin{aligned} \cos x &= \sqrt{1 - \sin^2 x} \\ &= \frac{\sqrt{24}}{5} = \frac{2\sqrt{6}}{5}\,. \\ \text{and} \quad \tan x &= \frac{\sin x}{\cos x} \\ &= \frac{1}{\sqrt{24}} = \frac{1}{2\sqrt{6}}\,. \end{aligned}$$

---

12. In a triangle labelled by its angles A, B, C, the small case letters a, b, c denote lengths opposite the corresponding angles. Determine all the sides and angles, and also the area, of the triangle for which $a = 5$, $b = 7$ and $C = 80°$.

– –

Solution:
Refer to the triangle labelled as in Fig.(2.5).

The Cosine Rule gives

$$\begin{aligned} c^2 &= a^2 + b^2 - 2ab\cos C \\ c &= \sqrt{5^2 + 7^2 - 2(5)(7)\cos 80^0} \\ &= \sqrt{74 - 12.16} = 7.86. \end{aligned}$$

Then using the Sine Rule,

$$\begin{aligned} \frac{5}{\sin A} &= \frac{c}{\sin C} \\ \sin A &= \frac{5\sin C}{c} = \frac{5\sin 80^0}{7.86} \\ \Rightarrow A &= \sin^{-1} 0.626 = 38.8^0. \end{aligned}$$

Finally, since the angles of a triangle add to 180^0, $B = 180^0 - 80^0 - 38.8^0 = 61.2^0$.

The area of triangle ABC is given by (see Eq.(2.1))

$$\begin{aligned} \text{Area} &= \frac{1}{2}ab\sin C \\ &= \frac{1}{2}(5)(7)\sin 80^0 \\ &= 17.2. \end{aligned}$$

---

13. A circle is inscribed in a regular heptagon of side one. What is the shortest distance from the centre of the circle to the heptagon?

- -

Solution:
The figure illustrates the situation. Since a heptagon has seven sides, the angle subtended at the centre by each side is $2\pi/7$. The shortest distance is perpendicular to a side and is labelled x.

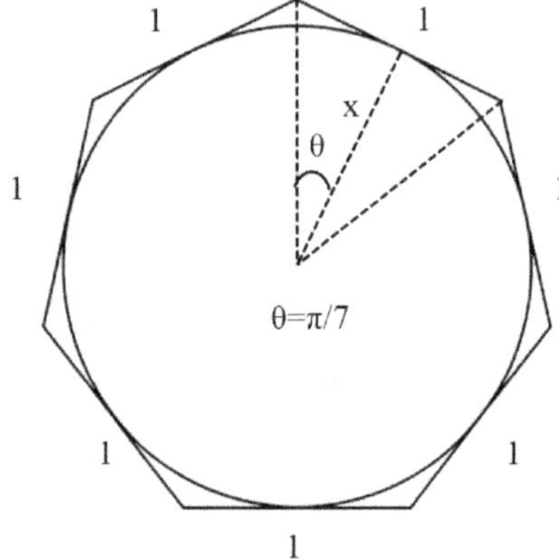

Figure 2.9: Figure of a circle inscribed in a regular heptagon of side one.

From the figure,

$$\begin{aligned} \tan\frac{\pi}{7} &= \frac{1/2}{x} \\ \Rightarrow x &= \frac{1/2}{\tan\pi/7} = 1.04. \end{aligned} \tag{2.7}$$

Hence the shortest distance from the center of the circle to the heptagon is $x = 1.04$.

---

14. From point B, the point C is due east, while A is at a bearing 020°. Also, $\angle BAC = 50°$. Draw a diagram, labelling all the angles and find the bearings of the following:

(a) A from C.

(b) C from A.

- -

Solutions:

(a) *Note: A bearing denotes a direction relative to North.* The diagram below illustrates the situation. Given the bearing of A, we deduce $\angle ABC = 90^0 - 20^0 = 70^0$. Then since $\angle BAC = 50°$, we get $\angle ACB = (180 - 70 - 50)^0 = 60^0$. Relative to point C, point A is 30^0 West of North.

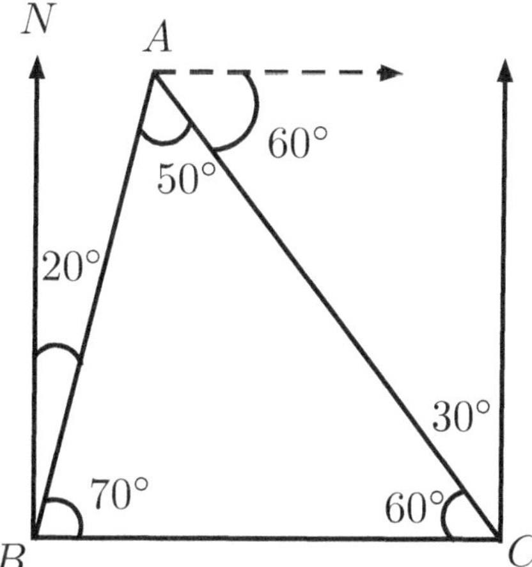

Figure 2.10: Figure for the bearings question.

Therefore, the bearing of A from C is $360^0 - 30^0 = 330^0$.

(b) Similarly, the bearing of C from A is $90^0 + 60^0 = 150^0$.

---

15. A right angled triangle is inscribed in a circle. The longest and shortest sides of the triangle are 10 and 5. Determine
(a) The diameter, D, of the circle.
(b) The length of the other side of the triangle.

\- -

Solutions:

(a) Refer to the figure below where O is the centre of the circle. By the Inscribed Angle Theorem (see Note below), $\angle AOB = 2\angle ACB = 180°$. Therefore the hypotenuse AB coincides with the diameter. Since by Pythagoras' theorem the hypotenuse is the longest side, $D = 10$.
(*Note: The Inscribed Angle Theorem. Let A,B and C be points located cyclically on the circumference of a circle of centre O. Then* $\angle AOC = 2\angle ABC$.)

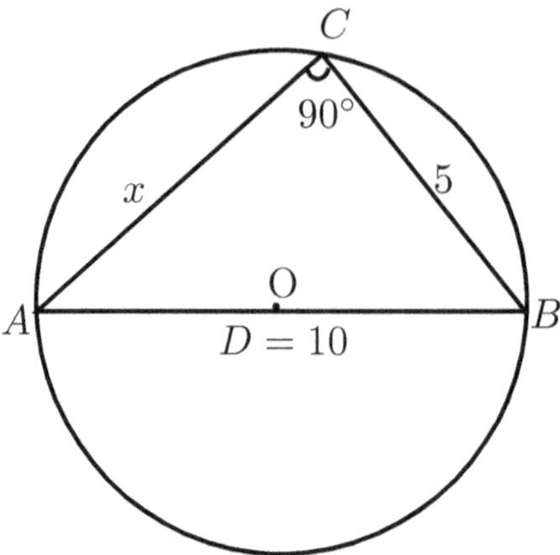

Figure 2.11: Exercise 15.

(b) By Pythagoras theorem, the length x of the other side of the triangle is determined by

$$\begin{aligned}(10)^2 &= 5^2 + x^2 \\ x &= \sqrt{(10)^2 - 5^2} = 5\sqrt{3}.\end{aligned}$$

---

16. The points A, B, C and D lie cyclically on the circumference of a circle with AC and BD being diameters and O the centre. If $\angle CAD = 25°$, find

(a) $\angle ABD$.
(b) $\angle BCA$.
(c) $\angle COD$.
(d) $\angle BAD$.

- -

Solutions:

(a) Since OA=OD, $\triangle AOD$ is isosceles. So $\angle CAD = \angle BDA = 25°$. Also, $\angle DAB$ is a right-angle since BD is the diameter. Thus, $\angle ABD = 90° - 25° = 65°$ as labelled in the figure.

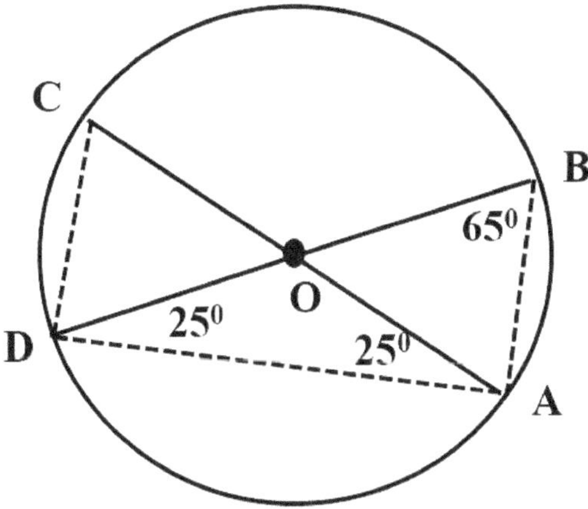

Figure 2.12: Example 16.

(b) $\angle BCA = \angle BDA = 25°$ as they are angles subtended by the same chord (on the same side).

(c) $\angle COD = 2\angle CAD = 50°$, by the Inscribed Angle Theorem (see Note in previous example).

(d) Since BD is the diameter, $\angle BAD = 90°$ (see also the previous worked example).

Did You Know?

Heron's formula for the area of a triangle with sides a, b, c is $A = \sqrt{s(s-a)(s-b)(s-c)}$ where s is half the perimeter.

Can you prove it?

2.5 Test Yourself

1. Find the range of values of x which satisfy $5 - 2x > 2 - x > 3x - 4$.

2. Find the range of values of x which satisfy $x^2 - x - 1 > 4$ and $3x^2 - 4 > x - 2$.

3. Find the equation of a straight line that passes through the point $(2, 3)$ and is parallel to the line $2x + 3y - 1 = 0$.

4. If in Worked Example (8), $b = a + 1$ and length $AB=$ length BC, determine a and b.

5. In a triangle labelled by its angles $A,\ B,\ C$, the small case letters $a,\ b,\ c$ denote lengths opposite the corresponding angles. Determine all the sides and angles if $A = 50°,\ B = 120°,\ c = 7$. Find also the area of the triangle.

6. For Worked Example (14), find the bearings of B from C and A respectively.

7. Challenge: A circle is inscribed inside a triangle with sides $3, 5$ and 6.

   (a) What is the area, T, of the triangle?

   (b) Find the radius, r, of the inscribed circle.

   (c) Find the radius, R, of the circle that circumscribes the triangle.

Did You Know?

The first 25 digits of π:

3.141 592 653 589 793 238 462 643 3....

2.6 Answers to Test

1. The common domain is $x < \frac{3}{2}$.

2. The common domain satisfying both conditions is $x < \frac{1-\sqrt{21}}{2}$ or $x > \frac{1+\sqrt{21}}{2}$.

3. $y = -2x/3 + 13/3$.

4. $(a, b) = (3, 4)$ or $(-3, -2)$.

5. $C = 10°$ and $a = 30.88$, $b = 34.91$. Area $= 93.6$.

6. $270°$ and $200°$.

7. *Hints:* For the area, use Eq.(2.1), getting the sin from the Cosine rule and $\sin\theta = \sqrt{1-\cos^2\theta}$ (This procedure essentially leads to Heron's formula).
   For r, connect each vertex of the triangle to the centre of the inscribed circle, then by adding the areas of the three smaller triangles, show that $r(a+b+c)/2 = T$ where a, b, c are the sides of the triangle.
   To get R, connect each vertex of the triangle to the centre of the circumscribing circle and show $a/\sin A = 2R$.

   Answers: $T = 2\sqrt{14}$. $r = 2\sqrt{14}/7$. $R = 45\sqrt{14}/56$.

Did You Know?

$$\frac{\pi}{4} = \sum_{n=0}^{\infty} \frac{(-1)^n}{2n+1} = 1 - \frac{1}{3} + \frac{1}{5} - \frac{1}{7} + \ldots$$

Chapter 3

Additional Algebra

3.1 Review: Exponents and Logarithms

Given the function $y = b^x$, the inverse relation is written $x = \log_b y$ where 'log' is short for the **logarithm** function, in this case to base b. That is,

$$y = b^x \Leftrightarrow x = \log_b y \,. \tag{3.1}$$

The choice $b = 10$ gives the **common logarithm**, while the **natural logarithm** has the irrational base $e = 2.718281....$ The exponential function e^x occurs in many mathematical laws describing natural phenomena as we will see in the practice questions.

The notation $\log_e$ is often abbreviated as ln, while $\log_{10}$ is sometimes written as lg.

Various identities for logarithms may be derived from those for exponents listed in Chapter 1. For $a, b > 0, a \neq 1, b \neq 1$ and $P, Q > 0$,

$$\log_b PQ = \log_b P + \log_b Q \,. \tag{3.2}$$

$$\log_b \frac{P}{Q} = \log_b P - \log_b Q \,. \tag{3.3}$$

$$\log_b P^c = c \log_b P \,. \tag{3.4}$$

$$\log_b P = \frac{\log_a P}{\log_a b} \,. \tag{3.5}$$

$$\log_b a = \frac{1}{\log_a b} \,. \tag{3.6}$$

The identity (3.5) is used for changing the bases of logarithms.

For $a > 1$ and $x > 0$, $y = \log_a x$ is an increasing function of x; it is positive for $x > 1$, negative for $x < 1$ and vanishes at $x = 1$.

Since $y = \log_a x$ is the inverse function to the function $y = a^x$, its graph may be obtained by reflecting the exponential curve about the line $y = x$.

3.1.1 Binomial Theorem

For a positive integer n, we have the **Binomial Theorem**,

$$\begin{aligned}(a+b)^n &= a^n + \binom{n}{1} a^{n-1} b + \binom{n}{2} a^{n-2} b^2 + \binom{n}{3} a^{n-3} b^3 + ... \\ &\quad + \binom{n}{r} a^{n-r} b^r + ... + \binom{n}{n-1} a b^{n-1} + b^n \,, \end{aligned} \tag{3.7}$$

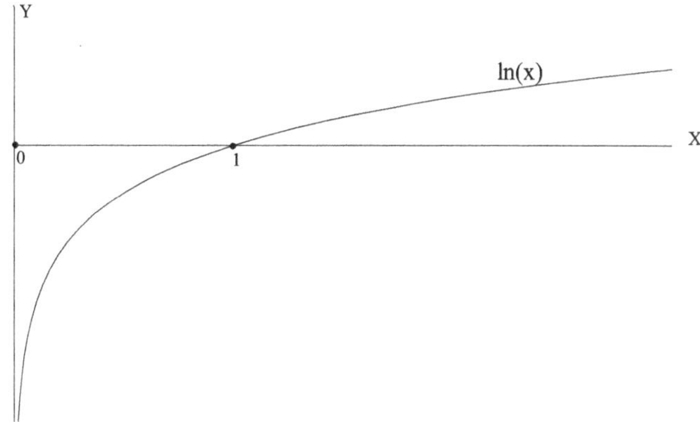

Figure 3.1: The natural logarithm.

where the binomial coefficient

$$^nC_r \equiv \binom{n}{r} = \frac{n!}{r!(n-r)!} \tag{3.8}$$

counts the number of ways of choosing r objects from n identical ones. The factorial symbol is defined by $n! = n \times (n-1) \times (n-2) \times .... \times 2 \times 1$ with $0! \equiv 1$.

Note that the $(r+1)$-th term in the above expansion is given by

$$\binom{n}{r} a^{n-r} b^r . \tag{3.9}$$

3.2 Review: Polynomials and Rational Functions

If $P(x)$ is a polynomial with $P(\alpha) = 0$, then $x = \alpha$ is a **root** of that equation and one may **factorise** $P(x) \equiv (x - \alpha)Q(x)$ where $Q(x)$ is a polynomial of one degree lower; this is the **Factorisation Theorem**. The process of factorisation can continue if there are more roots.

On the other hand, if $x = \alpha$ is not a root of $P(x)$ then one would obtain a remainder when $P(x)$ is divided by $(x - \alpha)$ using the process of long division, $P(x) \equiv (x - \alpha)Q(x) + R$ with $P(\alpha) = R$ the remainder: This summarises the **Remainder Theorem**.

One may obtain relations between the roots of a quadratic equation without explicitly solving for its roots. Denoting the two roots of the quadratic equation $Ax^2 + Bx + C = 0$ by α and β, leads to the two basic relations

$$\alpha + \beta = \frac{-B}{A} , \tag{3.10}$$

$$\alpha\beta = \frac{C}{A} . \tag{3.11}$$

A **rational function** is a ratio of two polynomials. In some applications it is convenient to re-write the ratio as the sum of rational functions whose denominators are the factors of the original denominator.

For a rational function $P(x)/Q(x)$, the first step in the decomposition is to use long division to reduce the degree of the numerator to below that of the denominator. Next, each factor of $(x-a)$ in $Q(x)$ would require a partial fraction $A/(x-a)$. If the factor is repeated in Q, for example $(x-a)^2$, then one uses two **partial fractions** $A_1/(x-a)$ and $A_2/(x-a)^2$ for that factor. If Q contains a term that cannot be factorised (using real numbers), for example x^2+x+1, then the partial fraction for that term is of the form $(Ax+B)/(x^2+x+1)$, that is, the numerator is one degree lower than the denominator.

3.3 Worked Examples

1. Simplify each of the following expressions, as much as possible, without using a calculator:

   (a) $\log_5 7 + 7\log_5 \dfrac{1}{49} + \log_5 \sqrt{35}$.

   (b) $\dfrac{5-2\sqrt{7}}{\sqrt{7}+3}$.

- -

Solutions:

(a)

$$\begin{aligned}
\log_5 7 + 7\log_5 \frac{1}{49} + \log_5 \sqrt{35} &= \log_5 7 + \log_5 (7^{-2})^7 + \log_5 (7\times 5)^{1/2} \\
&= \log_5 \left[7 \times (7^{-2})^7 \times (7\times 5)^{1/2}\right] \\
&= \log_5 \left[7 \times 7^{-14} \times 7^{1/2} \times 5^{1/2}\right] \\
&= \log_5 \left[7^{-12.5} \times 5^{0.5}\right] \\
&= -12.5\log_5 7 + 0.5\log_5 5 \\
&= \frac{1}{2} - \frac{25}{2}\log_5 7 \ ,
\end{aligned}$$

where we made use of $\log_5 5 = 1$ in the last step.

(b) The standard method is to multiply the numerator and denominator by the conjugate of the denominator: *Note:* $(a-b)(a+b) = a^2 - b^2$.

$$\begin{aligned}
\frac{5-2\sqrt{7}}{\sqrt{7}+3} &= \frac{5-2\sqrt{7}}{\sqrt{7}+3} \times \frac{\sqrt{7}-3}{\sqrt{7}-3} \\
&= \frac{11\sqrt{7}-29}{7-9} \\
&= \frac{29-11\sqrt{7}}{2} .
\end{aligned}$$

2. Find the values of $x > 0$ for which $\log_a x = a^x$, where $a > 3$ is a constant.

- -

Solution:
We note that $y = a^x > 1$ for $x > 0$ and $a > 1$. Also, a^x is an increasing function of x. A sketch using software shows that for $a > 3$, a^x lies above the line $y = x$.

Now, the reflection of $y = a^x$ about the line $y = x$ gives the function $x = a^y$ which lies below the line $y = x$ (for $a > 3$). Since $x = a^y$ implies $\log_a x = y$, therefore the function $\log_a x$ lies below the line $y = x$.

Hence the two curves a^x and $\log_a x$ do not intersect for $a > 3$: The solution set is empty.

---

3. Solve for x without using a calculator:
$\log_{10}(x+1) - \log_{10}(x-1) = 2 - \log_{10} 2$.

- -

Solution:

$$\begin{aligned}
\log_{10}(x+1) - \log_{10}(x-1) &= 2 - \log_{10} 2 \\
\log_{10}\frac{(x+1)}{(x-1)} + \log_{10} 2 &= 2 \\
\log_{10}\frac{2(x+1)}{(x-1)} &= 2 \\
\frac{2(x+1)}{(x-1)} &= 10^2 \\
x+1 &= \frac{100}{2}(x-1) \\
&= 50(x-1) \\
49x &= 51 \\
\therefore x &= \frac{51}{49}.
\end{aligned}$$

---

4. Find the first three terms, in increasing powers of x, in the expansion of $\left(x + \frac{1}{x}\right)^8$.

- -

Solution:
It is convenient to pull out the negative powers of x before applying the Binomial Theorem:

$$\begin{aligned}\left(x+\frac{1}{x}\right)^8 &= \frac{1}{x^8}\left(1+x^2\right)^8 \\ &= \frac{1}{x^8}\left[1+\binom{8}{1}(x^2)^1+\binom{8}{2}(x^2)^2+...\right] \\ &= \frac{1}{x^8}(1+8x^2+28x^4+...\,) \\ &\approx \frac{1}{x^8}+\frac{8}{x^6}+\frac{28}{x^4}.\end{aligned}$$

---

5. Re-arrange the following sequence in order, from the smallest to the largest, without using a calculator: $\log_4 3,\ \log_2 2\sqrt{3},\ 2\log_9 3$. Justify your answer.

- -

Solution: *Recall eq.(3.5).*
First, re-write the terms in base 2. We have $\log_4 3 = \dfrac{\log_2 3}{\log_2 4} = \dfrac{\log_2 3}{\log_2 2^2} = \dfrac{\log_2 3}{2} = \log_2 \sqrt{3}$.

Also, $2\log_9 3 = \dfrac{2\log_2 3}{\log_2 9} = \dfrac{2\log_2 3}{\log_2 3^2} = 1 = \log_2 2$.

Since $\sqrt{3} < \sqrt{4} = 2$ and $2 < 2\sqrt{3}$, so $\log_4 3 < 2\log_9 3 < \log_2 2\sqrt{3}$.

---

6. Solve the following equation for x to two decimal places:
$(\lg x)^2 + \lg(\sqrt{x}) - 0.1 = 0$.

- -

Solution:
Let $y = \lg x$. Thus

$$\begin{aligned} 0 &= (\lg x)^2 + \lg\sqrt{x} - 0.1 \\ &= y^2 + \frac{y}{2} - 0.1 \\ \Rightarrow y &= \frac{-0.5 \pm \sqrt{0.65}}{2} \\ &= 0.153 \text{ or } -0.653 \\ \therefore x &= 10^y \\ &= 1.42 \text{ or } 0.22.\end{aligned}$$

---

7. Solve the following pair of simultaneous equations:
$27^x\ 3^{5y} = 9$ and $\log_2(1-x) - 1 = \log_2 y$.

- -

Solution:
We first simplify each equation. For the first,

$$\begin{aligned} 27^x \cdot 3^{5y} &= 9 \\ 3^{3x} \cdot 3^{5y} &= 3^2 \\ 3^{3x+5y} &= 3^2 \\ \Rightarrow 3x + 5y &= 2\ . \end{aligned} \tag{3.12}$$

For the second equation,

$$\begin{aligned} \log_2 y &= \log_2(1-x) - 1 \\ \log_2 \frac{y}{1-x} &= -1 \\ \Rightarrow y &= \frac{1-x}{2}. \end{aligned} \tag{3.13}$$

The two linear simultaneous equations (3.12) and (3.13) can be solved easily to give $x = -1$ and $y = 1$.

---

8. Solve the following equation for x: $\sqrt{x-1} + \sqrt{x-4} = \sqrt{15-2x}$.

- -

Solution:
Square both sides and simplify to get

$$\begin{aligned} \sqrt{x^2 - 5x + 4} &= 10 - 2x \\ \Rightarrow x^2 - 5x + 4 &= 100 - 40x + 4x^2 \\ x &= \frac{35 \pm \sqrt{73}}{6}\ . \end{aligned}$$

However, we need to verify the solutions since we squared the original expression (see the book). Substitution into the initial equation shows that only $x = \dfrac{35 - \sqrt{73}}{6}$ is valid. (Note: When checking by substitution, you can use the approximate values of x to make a quick judgement.)

---

9. Using an appropriate sketch, show that the equation $e^{-x} - \sin x = 0$ has exactly two solutions between $0 \leq x \leq \pi$ and no solutions between $\pi \leq x \leq 2\pi$. Estimate the value of the solution in the range $0 \leq x \leq \pi/2$.

Solution:
We can solve the equation numerically by plotting both $f_1(x) = e^{-x}$ and $f_2(x) = \sin x$ on the same graph and looking for the intersection points.

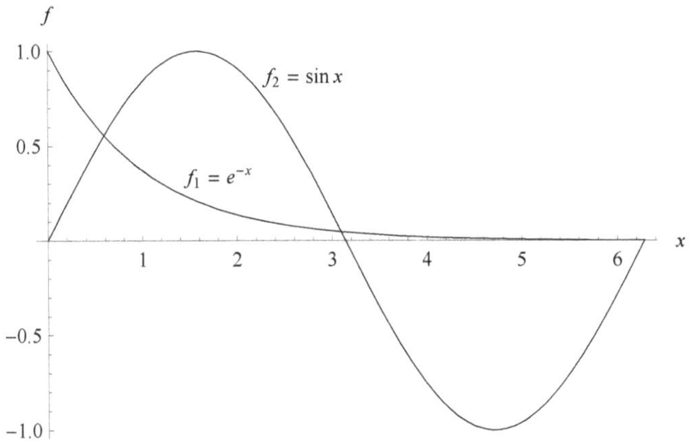

Figure 3.2: For example 9.

The sketch shows that there are exactly two solutions between $0 \leq x \leq \pi$ and no solution between $\pi \leq x \leq 2\pi$. From the figure, the solution in the range $0 \leq x \leq \pi/2$ is given by $x \approx 0.59$.

---

10. Solve the following equation for x:
$1 + \dfrac{3}{\sqrt{x}} - \dfrac{1}{x} = 0.$

Solution:

$$
\begin{aligned}
1 + \frac{3}{\sqrt{x}} - \frac{1}{x} &= 0 \\
\frac{x + 3\sqrt{x} - 1}{x} &= 0 \\
\Rightarrow x + 3\sqrt{x} - 1 &= 0.
\end{aligned}
$$

Next, we move the $3\sqrt{x}$ term to the right-hand side and square both sides,

$$\begin{aligned}(x-1)^2 &= (-3\sqrt{x})^2 \\ \Rightarrow x^2 - 11x + 1 &= 0 \\ x &= \frac{11 \pm \sqrt{121-4}}{2} \\ &= \frac{11 \pm 3\sqrt{13}}{2}.\end{aligned}$$

As in Example 8, we need to verify the answers. By substituting into the original equation, we see that only $x = (11 - 3\sqrt{13})/2$ is acceptable.

---

11. The roots of the quadratic equation $2x^2 - 7x + 1 = 0$ are α and β.

    (a) Determine the values of $\alpha + \beta$, $\alpha\beta$ and $1/\alpha + 1/\beta$ without solving the quadratic equation.

    (b) Solve the quadratic equation to determine α and β explicitly and verify your answers to part (a).

\- -

Solutions: *Recall Eq.(3.10) and Eq.(3.11).*

(a) For this problem,

$$\begin{aligned}\alpha + \beta &= \frac{7}{2}; \\ \alpha\beta &= \frac{1}{2}; \\ \frac{1}{\alpha} + \frac{1}{\beta} &= \frac{\alpha+\beta}{\alpha\beta} = \frac{7/2}{1/2} = 7.\end{aligned}$$

(b)

$$x = \frac{7 \pm \sqrt{49 - 4(2)}}{2(2)} = (7 \pm \sqrt{41})/4.$$

Let $\alpha = (7 + \sqrt{41})/4$ and $\beta = (7 - \sqrt{41})/4$. We leave the straight-forward verification to you.

---

12. If the roots of the quadratic equation $ax^2 + bx + c = 0$ are α and β, determine the roots of the equation $cx^2 + bx + a = 0$ in terms of α and β.

- -

Solution:

The solutions of $ax^2 + bx + c = 0$ are $\dfrac{-b \pm \sqrt{b^2 - 4ac}}{2a}$. Label the '+' solution α and the other β. Note that $\alpha\beta = c/a$. The solutions of $cx^2 + bx + a = 0$ are

$$\begin{aligned} \frac{-b \pm \sqrt{b^2 - 4ac}}{2c} &= \left(\frac{a}{c}\right) \frac{-b \pm \sqrt{b^2 - 4ac}}{2a}. \\ &= \frac{1}{\alpha\beta} \times (\alpha,\ \beta) \\ &= \frac{1}{\beta} \text{ and } \frac{1}{\alpha}\,. \end{aligned}$$

---

13. The cubic equation $f(x) = x^3 - ax^2 - x + 3$ has a root at $x = -1$.

    (a) Factorise $f(x)$.

    (b) Sketch the equation $y = f(x)$.

    (c) For what values of c will the curve in part (b) intersect the curve $x^3 - y - c = 0$?

- -

Solution:

(a) Given that the cubic equation $f(x) = x^3 - ax^2 - x + 3$ has a root at $x = -1$ implies $f(-1) = 0$. Therefore $-1 - a + 1 + 3 = 0 \Rightarrow a = 3$. Then long division gives,

$$\begin{array}{rl} & x^2 - 4x + 3 \\ \hline x+1) & x^3 - 3x^2 - x + 3 \\ & \underline{-(x^3 + x^2)} \\ & \quad -4x^2 - x + 3 \\ & \quad \underline{-(-4x^2 - 4x)} \\ & \qquad 3x + 3 \\ & \qquad \underline{3x + 3} \\ & \qquad\quad 0\,. \end{array}$$

So, $f(x) = (x + 1)(x^2 - 4x + 3)$ and the other roots of $f(x)$ are the roots of the quadratic equation. After factorising the quadratic part, we get
$f(x) = (x + 1)(x - 1)(x - 3)$.

(b) Sketch of the curve:

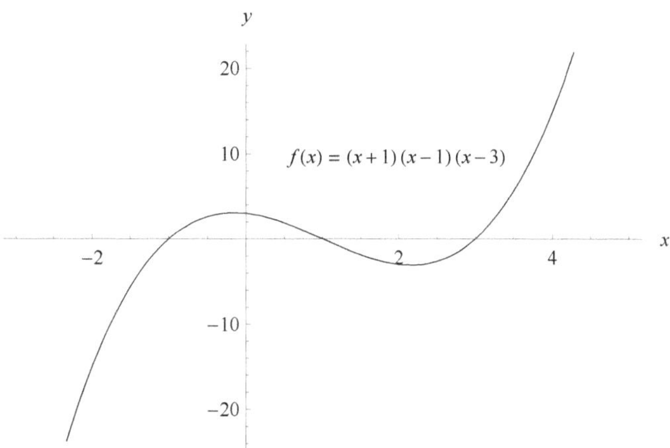

Figure 3.3: Plot of $f(x) = (x+1)(x-1)(x-3)$.

(c) We substitute $y = x^3 - c = 0$ into $y = f(x)$ to find the intersection points. We get

$$\begin{aligned} x^3 - c &= x^3 - 3x^2 - x + 3 \\ 0 &= 3x^2 + x - (c+3). \end{aligned}$$

Imposing the discriminant condition $B^2 - 4AC \geq 0$ to obtain real solutions of the quadratic equation, we get

$$\begin{aligned} 1^2 - 4(3)(-c-3) &\geq 0 \\ 12c + 37 &\geq 0 \\ \therefore\ c &\geq -37/12. \end{aligned}$$

---

14. If the roots of the quadratic equation $2x^2 - 7x + p = 0$ are α and β, with $\alpha > \beta$, determine the following in terms of p:

    (a) $\alpha - \beta$.

    (b) $(\beta/\alpha + \alpha/\beta)$.

- -

Solutions: *Recall Eq.(3.10) and Eq.(3.11).*
We have $\alpha + \beta = 7/2$ and $\alpha\beta = p/2$.

(a) Then

$$\begin{aligned} \alpha - \beta &= \sqrt{(\alpha-\beta)^2} \\ &= \sqrt{(\alpha+\beta)^2 - 4\alpha\beta} \\ &= \sqrt{(7/2)^2 - 4(p/2)} \\ &= \frac{\sqrt{49-8p}}{2}. \end{aligned}$$

We chose only the positive square-root solution since $\alpha > \beta$.

(b)

$$\begin{aligned}\frac{\beta}{\alpha}+\frac{\alpha}{\beta} &= \frac{\alpha^2+\beta^2}{\alpha\beta}\\ &= \frac{(\alpha+\beta)^2-2\alpha\beta}{\alpha\beta}\\ &= \frac{(7/2)^2-2(p/2)}{p/2}\\ &= \frac{49}{2p}-2.\end{aligned}$$

---

15. A teacher tells her class that the quartic polynomial $P(x) = x^4 + ax^3 + bx^2 + 3$ is identical to the expression $S(x) = x(x+1)(x-2)Q(x) + c + 1$ written by a student, where $Q(x)$ is a polynomial and a, b, c are constants.

  (a) Determine the constants a, b and c.

  (b) Hence find an expression for $Q(x)$.

  (c) What is the remainder when the polynomial $P(x)$ is divided by $x - 1$?

- -

Solutions:

(a) Given that $P(x) = x^4+ax^3+bx^2+3$ is identical to $S(x) = x(x+1)(x-2)Q(x)+c+1$, we can evaluate both expressions for particular values of x and compare the results. When $x = 0$ we have $P(0) = 3; S(0) = c + 1$, thus $c = 2$. Next by letting $x = -1$, $P(-1) = S(-1) \Rightarrow a - b = 1$. Also, $P(2) = S(2) \Rightarrow 2a + b = -4$.
So we have the two simultaneous equations $a - b = 1$ and $2a + b = -4$ which can be easily solved to give $a = -1, b = -2$.

(b) Setting $P(x) \equiv S(x)$ gives

$$\begin{aligned}x^4 - x^3 - 2x^2 + 3 &= x(x+1)(x-2)Q(x) + 3\\ x^3 - x^2 - 2x &= (x+1)(x-2)Q(x)\\ x(x^2 - x - 2) &= (x^2 - x - 2)Q(x)\\ \Rightarrow Q(x) &= x\,. \end{aligned} \tag{3.14}$$

(c) By the remainder theorem, the remainder is

$$P(1) = (1)^4 - (1)^3 - 2(1)^2 + 3 = 1.$$

---

16. Re-write the following in terms of partial fractions: $\dfrac{5x^2+2x+1}{x^2+2x+1}$.

- -

Solution:
After a long division, we get

$$\frac{5x^2+2x+1}{x^2+2x+1} = 5 - \frac{8x+4}{x^2+2x+1}$$

Since $(x^2+2x+1) = (x+1)^2$, the partial fractions will be of the form

$$\frac{-(8x+4)}{x^2+2x+1} \equiv \frac{A}{x+1} + \frac{B}{(x+1)^2} . \tag{3.15}$$

Therefore

$$A(x+1)+B = -(8x+4) .$$

Substituting $x=-1$ into Eq.(3.15) gives $B=4$ and setting $x=0$ gives $A=-8$. Thus

$$\frac{5x^2+2x+1}{x^2+2x+1} = 5 - \frac{8}{x+1} + \frac{4}{(x+1)^2} .$$

---

17. In each case below, solve the simultaneous equations for x and y:

   (a) $y^2 = 2xy + 3x + 2$ and $y = |x-1|$.

   (b) $2y - 5\sqrt{x} = 1$ and $2x - 5y = 3$.

- -

Solutions:

(a) We consider two separate cases

$$y = |x-1| = \begin{cases} x-1 & \text{if } x \geq 1 \\ 1-x & \text{if } x < 1 \end{cases}$$

For the case $y = x-1$, substitution into the other equation gives $(x-1)^2 = 2x(x-1)+3x+2$. This leads to the quadratic equation $x^2+3x+1=0$ with roots $x = (-3\pm\sqrt{5})/2$ that are negative. This is an invalid solution since $x \geq 1$ for this case.

For the second case $y = 1-x$, we get $3x^2-7x-1=0$ which gives the valid solution $x = (7-\sqrt{61})/6$. (The other root is ignored since $x<1$). The corresponding y value is $y = 1-x = (\sqrt{61}-1)/6$.

(b) Re-write the first equation as

$$\begin{aligned} 5\sqrt{x} &= 2y-1 \qquad (3.16)\\ x &= \left(\frac{2y-1}{5}\right)^2. \end{aligned}$$

From (3.16) we see that $x \geq 0$ and also $y \geq 1/2$. Next, we replace x in the second equation and solve the resulting quadratic equation for y. Explicitly,

$$\begin{aligned} 3 &= 2\left(\frac{2y-1}{5}\right)^2 - 5y \\ 0 &= 8y^2 - 133y - 73 \\ \Rightarrow y &= \frac{133 \pm \sqrt{133^2 - 4(8)(-73)}}{2(8)} \\ &= \frac{133 \pm 15\sqrt{89}}{16}. \end{aligned}$$

Since $y \geq 1/2$, we only accept $y = (133 + 15\sqrt{89})/16$. The corresponding x is

$$\begin{aligned} x &= \frac{3 + 5 \times \frac{133+15\sqrt{89}}{16}}{2} \\ &= \frac{713 + 75\sqrt{89}}{32}. \end{aligned}$$

Did You Know?

10^{100} is called a "googol".

$10^{10^{100}} = 10^{(\text{googol})}$ is a "googolplex".

3.4 Test Yourself

1. Show that if $a,\ b > 1$, then $\log_a b > \log_b a$ implies $b > a$.

2. Solve the following equation for x without using a calculator:
$\dfrac{\log_2 x^2}{2\log_x 2} = 9.$

3. Find the term independent of x in the expansion of the expression in Worked Example 4 of Chapter 2.

4. The roots of the quadratic equation $ax^2 + bx + c = 0$ are p and q, while the roots of the quadratic equation $bx^2 + cx + a = 0$ are r and s.
Express $r + s$ and rs in terms of p and q.

5. The roots of the quadratic equation $ax^2 + bx + c = 0$ are p and q. The roots of the cubic equation $f(x) = 0$ are $p + q$, pq and $1/p + 1/q$.
If $f(0) = -1$, express $f(x)$ explicitly in terms of a, b, c.

6. In each case below, solve the simultaneous equations for x and y:

   (a) $\dfrac{1}{y - \sqrt{x}} = \dfrac{1}{\sqrt{x} + 2y} = -1.$

   (b) $x + y = 2xy + 3$ and $xy = 4$.

7. Find the values of x which satisfy the following: $2x^2 - 3x + 3 > 2 - x > 3x - 4$.

8. Challenge: Re-write the following expression in the form $a + \sqrt{b}$ where a, b are rational numbers:

$$\frac{\sqrt{7+4\sqrt{3}}+2}{\sqrt{7+4\sqrt{3}}-2}\,. \tag{3.17}$$

Did You Know?

$$\lim_{n\to\infty}\left(1+\frac{1}{n}\right)^n = e,$$

where e is Euler's constant.

3.5 Answers to Test

1. Using Eq.(3.6) we get $(\log_a b)^2 > 1 \Rightarrow \log_a b > 1$ or $\log_a b < -1$. The first condition implies $b > a$. The second condition gives $b < 1/a$, but since $a > 1$ this would imply $b < 1$ which contradicts the given condition $b > 1$. Hence the solution is $b > a$.

2. 8 and 1/8.

3. 70.

4. $r + s = pq/(p+q)$, $rs = -1/(p+q)$.

5. $\dfrac{a^2}{b^2}(x - c/a)(x + b/a)(x + b/c)$.

6. (a) $x = 1/9,\ y = -2/3$.

   (b) $x = \dfrac{11 \pm \sqrt{105}}{2},\ y = \dfrac{11 \mp \sqrt{105}}{2}$.

7. $x < 3/2$.

8. *Hint: Simplify* $\sqrt{7 + 4\sqrt{3}}$ *first by expressing it as* $a + b\sqrt{3}$, *where* a, b *are rational numbers.*

   Answer: $1 + \dfrac{4\sqrt{3}}{3}$.

Did You Know?

It is mathematically possible to construct a three-dimensional figure that has finite volume but infinite surface area.

Can you figure out one way to do it?

Chapter 4

Additional Geometry and Trigonometry

4.1 Review: Trigonometry

1. Some identities:

$$\sin^2 A + \cos^2 A = 1 \ . \tag{4.1}$$

$$1 + \tan^2 A = \sec^2 A \ . \tag{4.2}$$

$$\sin(-A) = -\sin A \ . \tag{4.3}$$

$$\cos(-A) = \cos A \ . \tag{4.4}$$

$$\tan(-A) = -\tan A \ . \tag{4.5}$$

2. Addition Formulae:

$$\sin(A \pm B) = \sin A \ \cos B \pm \cos A \ \sin B \ . \tag{4.6}$$

$$\cos(C \pm D) = \cos C \ \cos D \mp \sin C \ \sin D \ . \tag{4.7}$$

$$\tan(A \pm B) = \frac{\tan A \pm \tan B}{1 \mp \tan A \ \tan B} \ . \tag{4.8}$$

3. Special Cases of the Addition Formulae:

$$\sin(90^\circ \pm A) = \cos A \ . \tag{4.9}$$

$$\sin(180^\circ \pm A) = \mp \sin A \ . \tag{4.10}$$

4. Double Angle Formulae:

$$\sin 2A = 2\sin A \cos A \ . \tag{4.11}$$

$$\cos 2A = 2\cos^2 A - 1 = 1 - 2\sin^2 A = \cos^2 A - \sin^2 A \ . \tag{4.12}$$

$$\tan 2A = \frac{2\tan A}{1 - \tan^2 A} \ . \tag{4.13}$$

5. Factor Formulae:

$$\sin A \pm \sin B = 2\sin\frac{A \pm B}{2}\cos\frac{A \mp B}{2} \ . \tag{4.14}$$

$$\cos A + \cos B = 2\cos\frac{A + B}{2}\cos\frac{A - B}{2} \ . \tag{4.15}$$

$$\cos A - \cos B = -2\sin\frac{A + B}{2}\sin\frac{A - B}{2} \ . \tag{4.16}$$

6. R-Formulae:

$$a\cos\theta \pm b\sin\theta = R\cos(\theta \mp \alpha)\ ; \tag{4.17}$$
$$a\sin\theta \pm b\cos\theta = R\sin(\theta \pm \alpha)\ , \tag{4.18}$$

with $R = \sqrt{a^2+b^2}$, $\tan\alpha = b/a$.

4.2 Review: Coordinate Geometry

The equation for a straight line may be written in the form

$$y = mx + c \tag{4.19}$$

where m is the slope and c the intercept on the y-axis. If another line is perpendicular to (4.19), its slope must be $-1/m$.

The mid-point of a line joining two points (x_1, y_1) and (x_2, y_2) is given by the formula $\left(\frac{x_1+x_2}{2}, \frac{y_1+y_2}{2}\right)$.

The points on a circle are equidistant from a centre. If (x_0, y_0) are the coordinates of the centre and (x, y) a point on the circle of radius r, then

$$(x-x_0)^2 + (y-y_0)^2 = r^2\,, \tag{4.20}$$

which when expanded may be written in the form

$$x^2 + y^2 - 2ax - 2by + c = 0\,, \tag{4.21}$$

for some constants a, b, c. Conversely, given an equation in the form (4.21) one can complete the squares to get $(x-a)^2+(y-b)^2 = a^2+b^2-c$; if $c < a^2+b^2$ then (4.21) represents a circle with centre (a, b) and radius $R = \sqrt{a^2+b^2-c}$.

A tangent to a circle at any point P on its circumference is perpendicular to the radial line OP where O is the centre of the circle. Therefore the **normal** to the circle at P lies along OP.

Given three points $A(x_1, y_1)$, $B(x_2, y_2)$ and $C(x_3, y_3)$, with their relative order being anti-clockwise, the area of the triangle ABC may be obtained from the formula

$$A = \frac{1}{2}\left|x_1(y_2-y_3) + x_2(y_3-y_1) + x_3(y_1-y_2)\right|\,. \tag{4.22}$$

Notice the symmetry in the expression, in particular the cyclic $1 \to 2 \to 3 \to 1$ etc. labelling of indices. The formula is often presented in the "matrix" form

$$A = \frac{1}{2}\begin{vmatrix} x_1 & x_2 & x_3 & x_1 \\ y_1 & y_2 & y_3 & y_1 \end{vmatrix} \tag{4.23}$$

and the terms are generated as follows. First, start at the left of the top row and multiply each term in the top row by a term one step to the right in the bottom row, adding the pieces: $x_1y_2 + x_2y_3 + x_3y_1$. Then start at the right of the top row and multiply each term in the top row by a term one step to the left in the bottom row, adding the pieces: $x_1y_3 + x_3y_2 + x_2y_1$. Finally, subtract the two contributions and include the overall 1/2 to get

$$A = \frac{1}{2}(x_1y_2 + x_2y_3 + x_3y_1 - x_1y_3 - x_3y_2 - x_2y_1)\,, \tag{4.24}$$

which is the same as (4.22).

4.3 Review: Plane Geometry

Inscribed Angle Theorem: Let A and B be two points on the circumference of a circle. The angle subtended by A and B at the centre of the circle is twice that subtended at a point C on the circumference, see Fig.(4.1). This implies that two angles subtended by the same chord, on its same side, must be equal.

Tangent-Chord Theorem (Alternate Segment Theorem): Let the triangle ABC be inscribed in a circle and a tangent drawn at point A. Let D be another point on the tangent line such that D and C are on opposite sides of the line AB. Then $\angle DAB = \angle BCA$. See Fig.(4.2).

Intersecting Chords Theorem: Let A, B, C, and D be points on the circumference of a circle and X the point of intersection of the lines AC and BD. Then the triangle ABX is similar to triangle DCX. See Fig.(4.3). This implies $AX \cdot CX = BX \cdot DX$.

Tangent-Secant Theorem: Let A, B and C be points on the circumference of a circle and let the tangent at A meet the line CB produced at D. Then $(DA)^2 = DB \times DC$. See Fig.(4.4).

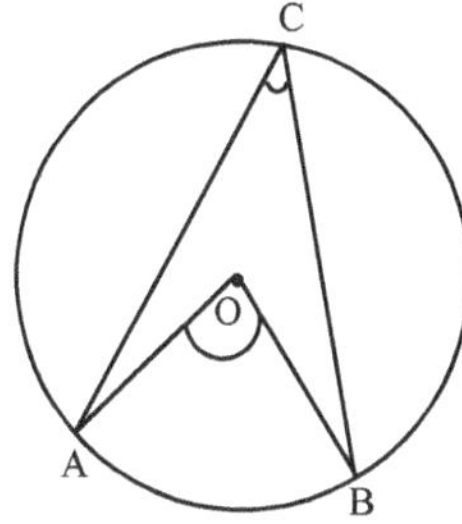

Figure 4.1: Figure illustrating the Inscribed Angle Theorem. Note that the relevant angle at the centre is on the side marked by the letter O.

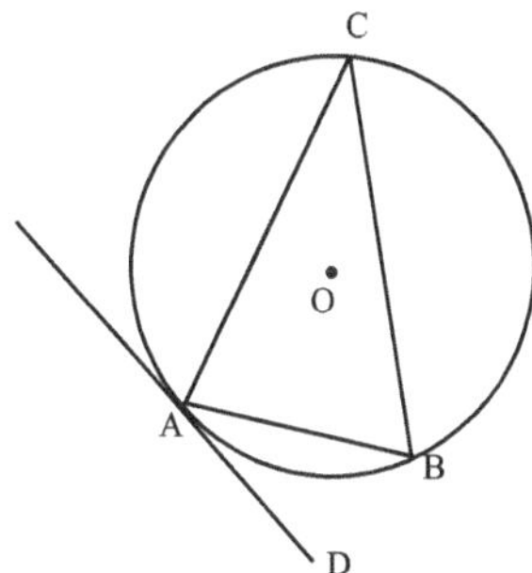

Figure 4.2: Figure illustrating the Tangent-Chord Theorem.

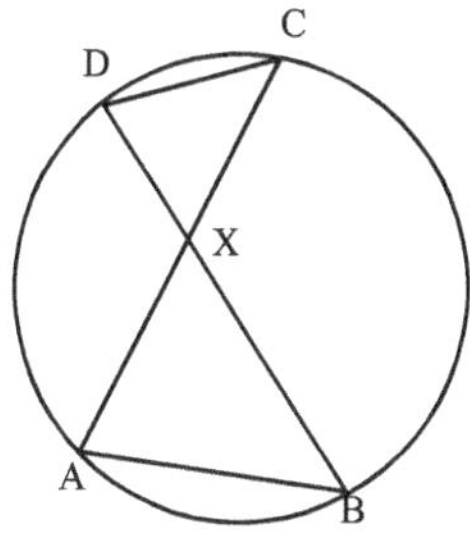

Figure 4.3: Figure illustrating the Intersecting Chords Theorem: Triangle ABX is similar to triangle DCX.

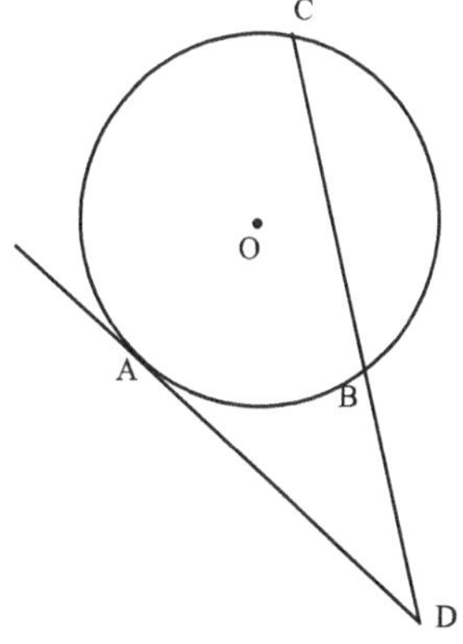

Figure 4.4: Figure illustrating the Tangent-Secant Theorem

4.4 Worked Examples

1. Re-arrange the following sequence in order, from the smallest to the largest, without using a calculator. Justify your answer.
   $\tan(55^\circ)$, $\sin(35^\circ)$, $\cos(35^\circ)$.

Solution:
Recall: $\sin\theta$ *is an increasing function for* $\theta \in [0, \pi/2]$ *and* $1 \geq \sin\theta \geq 0$.
To arrange the sequence in order, we compare the other terms to $\sin 35^\circ$. We have

$$\begin{aligned} \cos 35^\circ &= \sin(90^0 - 35^\circ) = \sin 55^\circ > \sin 35^\circ \ . \\ \tan 55^\circ &= \frac{\sin 55^\circ}{\cos 55^\circ} \\ &= \frac{\sin 55^\circ}{\sin 35^\circ} > 1 \ , \end{aligned} \tag{4.25}$$

where we used $\cos 55^\circ = \sin(90^\circ - 55^\circ) = \sin 35^\circ$. Hence $\sin 35^\circ < \cos 35^\circ < \tan 55^\circ$.

---

2. Given $\sin x = 1/5$, and $0 < x < \pi/2$, express exactly the values of

   (a) $\cos 2x$

   (b) $\sin(x + \pi/3)$.

Solutions:
First, using Eq.(4.1), we obtain $\cos x = \sqrt{1 - \sin^2 x} = 2\sqrt{6}/5$.

(a) Then,

$$\begin{aligned} \cos 2x &= 2\cos^2 x - 1 \\ &= 2\left(\frac{2\sqrt{6}}{5}\right)^2 - 1 \\ &= \frac{23}{25} \ . \end{aligned} \tag{4.26}$$

(b) Use the Addition Formula together with the exact values $\cos\pi/3 = 1/2$ and $\sin\pi/3 = \sqrt{3}/2$:

$$\begin{aligned}
\sin\left(x+\frac{\pi}{3}\right) &= \sin x\ \cos\frac{\pi}{3} + \cos x\ \sin\frac{\pi}{3} \\
&= \frac{1}{5}\times\frac{1}{2} + \frac{2\sqrt{6}}{5}\times\frac{\sqrt{3}}{2} \\
&= \frac{1+2\sqrt{18}}{10} \\
&= \frac{1+6\sqrt{2}}{10}\,.
\end{aligned} \tag{4.27}$$

---

3. Find the smallest positive x which solves the following equation:
$1 + \sin x = 3\cos^2 x$.

- -

Solution:

$$\begin{aligned}
1+\sin x &= 3\cos^2 x \\
&= 3(1-\sin^2 x) \\
\Rightarrow 0 &= 3\sin^2 x + \sin x - 2 \\
\sin x &= \frac{-1\pm\sqrt{1-4(3)(-2)}}{2(3)} \\
&= \frac{-1\pm 5}{6} = 2/3 \text{ or } -1.
\end{aligned}$$

When $\sin x = 2/3$, we have the angle $\sin^{-1}(2/3) = 0.73$ rad. On the other hand, when $\sin x = -1 \Rightarrow\ x = 3\pi/2 = 4.71$ rad. So the smallest positive x is $x = 0.73$ rad.

---

4. Given that $\tan A = 1/2$ and that A is in the first quadrant, express the values of each of the following without using a calculator:

   (a) $\sin A$.

   (b) $\cos(A/2)$.

- -

Solutions:

(a) You can use a right-angled triangle and Pythagoras' theorem to get

$$\sin A = \frac{1}{\sqrt{2^2+1^2}} = \frac{1}{\sqrt{5}}.$$

(b) First obtain $\cos A = 2/\sqrt{5}$ as in part 9(a). Then from the double-angle formula,

$$\begin{aligned}\cos(A/2) &= \sqrt{\frac{\cos A + 1}{2}} \\ &= \sqrt{\frac{5+2\sqrt{5}}{10}}.\end{aligned}$$

---

5. If A is in the first quadrant and B in the second, with $\sin A = 0.2$ and $\cos B = -0.3$, find the value of $\sin B + \cos A$ correct to two decimal places. .

- -

Solution:
Given $\sin A = 0.2$, with A in the first quadrant, and $\cos B = -0.3$, with B in the second quadrant, we have $\cos A = +\sqrt{1-(0.2)^2} = \sqrt{0.96}$ and $\sin B = +\sqrt{1-(-0.3)^2} = \sqrt{0.91}$. Then

$$\begin{aligned}\sin B + \cos A &= \sqrt{0.91} + \sqrt{0.96} \\ &\approx 1.93.\end{aligned}$$

---

6. Given the function $g(x) = a\sin^2 x - b\cos x + cx$ with $g(2) = 3$ and $g(-2) = 5$,
(a) Determine the exact value of c.
(b) If also $g(1) = 1$, determine a and b to two decimal places.

- -

Solutions:

7. (a) On substitution,

$$3 = g(2) = a\sin^2 2 - b\cos 2 + 2c\ .$$

and

$$\begin{aligned}5 &= g(-2) = a\sin^2(-2) - b\cos(-2) - 2c \\ &= a\sin^2(2) - b\cos(2) - 2c\ .\end{aligned}$$

Taking the difference between the two equations, we get $4c = -2 \Rightarrow c = -\frac{1}{2}$.

(b) $g(1) = 1 = a\sin^2 1 - b\cos 1 - \frac{1}{2}$. This allows us to write $b = \dfrac{a\sin^2 1 - 3/2}{\cos 1}$ and thus (from part a),

$$\begin{aligned} 3 &= a\sin^2 2 - \left(\frac{a\sin^2 1 - 3/2}{\cos 1}\right)\cos 2 - 1 \\ a &= \frac{4 - 3\cos 2/(2\cos 1)}{\sin^2 2 - \sin^2 1\cos 2/\cos 1} \\ &= 3.76\ . \end{aligned}$$

Then by direct substitution,

$$b = \frac{3.76\sin^2 1 - 3/2}{\cos 1} \approx 2.15.$$

---

8. In the triangle ABC, $a/b = 2$, $\sin B/\sin C = 0.898$ and $c = 5.57$. Find, to one decimal place,

   (a) The value of the smallest angle in the triangle.

   (b) The length of the longest side.

\- -

Solutions:

(a) From the sine rule, $a/\sin A = b/\sin B = c/\sin C \Rightarrow b = 0.898c = 5.002$. Also, $a = 2b = 2 \times 5.002 = 10.004$. From the cosine rule,

$$\begin{aligned} C &= \cos^{-1}\left(\frac{a^2 + b^2 - c^2}{2ab}\right) \\ &= \cos^{-1}\left(\frac{94.075}{100.080}\right) = 19.95^0 \approx 20^0\ . \end{aligned}$$

Then

$$\begin{aligned} B &= \sin^{-1}\left(\frac{b}{c}\sin C\right) \\ &= \sin^{-1}\left(0.898\sin 20^0\right) \approx 17.9^0\ , \end{aligned}$$

and $A = 180^0 - (20^0 + 17.9^0) = 142.1^0$. Thus, the smallest angle in the triangle is $\angle B = 17.9^0$. (Note: The smallest angle must be opposite the shortest side. So we could also have just evaluated B after deducing all the sides.)

(b) The length of the longest side is $a = 10$.

---

9. An equilateral triangle is inscribed in a circle which is itself inscribed in a unit square. What is the area of the triangle?

Solutions:
The radius is $r = \frac{1}{2}$ since the the circle is inscribed in a unit square. Let x be half of a side of the triangle. Then, from the figure below,

$$\cos 30^\circ = \frac{x}{r} \Rightarrow x = \frac{\sqrt{3}}{2} \times \frac{1}{2} = \frac{\sqrt{3}}{4}.$$

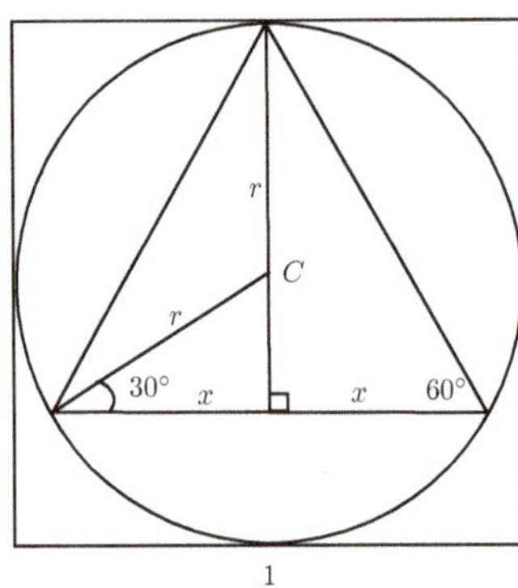

Figure 4.5: Example 9.

Thus, the area of the triangle is

$$\begin{aligned} A &= \frac{1}{2}(2x)(2x)\sin 60^\circ \\ &= \frac{3\sqrt{3}}{16}. \end{aligned}$$

---

10. An engineer is asked to design a roller coaster track which has a sinusoidal shape, $y(x) = A + B\cos(kx + C)$ where x is the straight line distance as measured along the ground and y the height of the track above ground. The requirements are: The lowest and highest points of the track should be respectively 20m and 30m above ground; the distance between the peaks of the track should be 50m; the start of the track at $x = 0$ should be 25 above ground. Can you help the engineer determine the constants A, B, k, C with $C < \pi$?

- -

Solutions:
Given $y(x) = A + B\cos(kx + C)$. Since the cosine function is bounded by $-1 \leq \cos x \leq 1$ for all x, and also given that the maximum (minimum) values of y are 30m (20m), we have

$$\begin{aligned} y_{\max} &= 30 = A + B \ , \\ y_{\min} &= 20 = A - B \ , \end{aligned}$$

which imply $A = 25$ and $B = 5$. The function is thus $y = 25 + 5\cos(kx + C)$. At the start of the track $x = 0, y = 25$, we have $\cos C = 0 \Rightarrow C = \pi/2$. At a peak, $y = 30$, so

$$\begin{aligned} 30 &= 25 + 5\cos\left(kx + \frac{\pi}{2}\right) \\ 1 &= \cos\left(kx + \frac{\pi}{2}\right) \\ \Rightarrow kx + \frac{\pi}{2} &= 0, 2\pi, 4\pi, \ldots \end{aligned}$$

Let $\Delta x = 50$ be the distance between two peaks. Then

$$\begin{aligned} k(\Delta x) &= 2\pi \\ k &= \frac{2\pi}{(\Delta x)} = \frac{2\pi}{50} \ . \end{aligned}$$

Hence $y(x) = 25 + 5\cos\left(\frac{2\pi x}{50} + \frac{\pi}{2}\right)$.

---

11. Determine how the number of intersection points of the line $y = mx - 2$ with the circle $(x-2)^2 + (y-1)^2 = 9$ depends on the parameter m.

- -

Solution:
To find the intersection points, substitute the straight line equation into the equation for the circle to get:

$$\begin{aligned} (x-2)^2 + (mx-3)^2 &= 9 \\ \Rightarrow (m^2+1)x^2 - (4+6m)x + 4 &= 0 \ . \end{aligned}$$

The discriminant of the above quadratic equation in x is $\Delta = 4m(5m + 12)$. Clearly there will be no real solutions for $\Delta < 0$; that is, when $-12/5 < m < 0$ the line does not intersect the circle.

For $\Delta = 0$, that is, when $m = 0$ or $m = -12/5$, there is one solution each, corresponding to the line being at a tangent to the circle at two possible points.

Finally, for $\Delta > 0$, that is, when $m > 0$ or $m < -12/5$, the line intersects the circle at two points.

(Note: A different method of solution is presented in the *Solutions Manual: Integrated Mathematics for Explorers, by Chee Leong Ching and Sun Jie.*)

---

12. A particle moves in the plane in such a way that both its x and y coordinates are oscillatory: $x = A \sin \omega t$ and $y = B \cos \omega t$, where t is the time variable and $A,\ B,\ \omega$ are constants. Eliminate the time variable to find the trajectory of the particle described by an equation relating x and y.

- -

Solution:
We write $\dfrac{x}{A} = \sin \omega t$ and $\dfrac{y}{B} = \cos \omega t$, and use the trigonometry identity

$$\begin{aligned} \sin^2 \omega t + \cos^2 \omega t &= 1 \\ \Rightarrow \frac{x^2}{A^2} + \frac{y^2}{B^2} &= 1\,. \end{aligned}$$

(Note: This is the equation of an ellipse, which becomes a circle when $A = B$).

---

13. The y-axis is tangent to a circle C. On the circumference of C, the point with the largest y-coordinate is $P(3, 10)$. Determine, with the help of a sketch,

(a) The equation of the circle.

(b) The point on the circle which has the largest value for its x-coordinate.

(c) The equation of a circle C_2 obtained by first reflecting C about the x-axis and then about the y-axis.

- -

Solutions:

(a) See the figure below. Given that the largest y-coordinate is $P(3, 10)$ and the y-axis is tangent to the curve C, this implies that the radius is $r = 3$ and the centre of the circle is at $(3, 7)$. Thus the equation of the circle is $(x-3)^2 + (y-7)^2 = 3^2$.

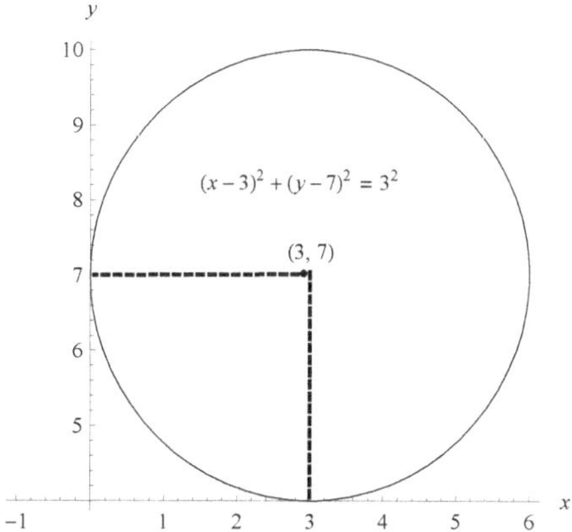

Figure 4.6: The plot of $(x-3)^2 + (y-7)^2 = 3^2$.

(b) The point with the largest x-coordinate is $(6, 7)$.

(c) The reflections produce circles with the same radius, 3, but different centres. The centre first moves from $(3, 7)$ to $(3, -7)$, then to $(-3, -7)$. Therefore the final circle has equation $(x+3)^2 + (y+7)^2 = 3^2$.

---

14. In analysing the data of a laboratory experiment, a student is asked to plot the variable $\ln y$ against x^2. The data seem to fit a straight line which makes an angle of 30° with the horizontal axis. The point $(x, y) = (0, 2)$ lies on the line. Determine the relationship between y and x.

– –

Solution:

Plotting $\ln y$ against x^2, the straight line with slope m is $\ln y = mx^2 + c$. This line forms an angle of 30° with the horizontal axis, therefore $m = \tan 30° = \frac{\sqrt{3}}{3}$. Also, since $(x, y) = (0, 2)$ is a point on the line,

$$\begin{aligned} \frac{\ln y - \ln 2}{x^2 - 0} &= \frac{\sqrt{3}}{3} \\ \ln \frac{y}{2} &= \frac{x^2}{\sqrt{3}} \\ \Rightarrow y &= 2e^{x^2/\sqrt{3}}. \end{aligned}$$

---

15. The curve $\mathcal{C}$ with equation $(y+1)(x+3) = 2$ intercepts the line $2x + 3y - 1 = 0$ at two points P and Q. Determine

    (a) The coordinates of the points P and Q.

    (b) The length of the line PQ.

    (c) The area of the triangle OPQ where $O = (0,0)$ is the origin.

    (d) The shortest distance from O to PQ.

- -

Solutions:

(a) We need to solve the simultaneous equations by substitution. From $2x + 3y - 1 = 0 \Rightarrow y = (1-2x)/3$, so we have

$$\begin{aligned} 2 &= (\frac{1-2x}{3} + 1)(x+3) \\ \Rightarrow 6 &= (4-2x)(x+3) \\ 0 &= x^2 + x - 3 \\ \Rightarrow x &= \frac{-1 \pm \sqrt{13}}{2} \end{aligned}$$

Correspondingly, $y = \frac{2 \mp \sqrt{13}}{3}$. The points are $P(\frac{-1+\sqrt{13}}{2}, \frac{2-\sqrt{13}}{3})$ and $Q(\frac{-1-\sqrt{13}}{2}, \frac{2+\sqrt{13}}{3})$.

(b) The length between any two points on the graph is given by $D = \sqrt{(x_2 - x_1)^2 + (y_2 - y_1)^2}$. For P and Q,

$$\begin{aligned} D_{PQ} &= \sqrt{(-\sqrt{13})^2 + (2\sqrt{13}/3)^2} \\ &= \frac{13}{3}. \end{aligned}$$

(c) The area of OPQ is

$$\begin{aligned} \text{Area} &= \frac{1}{2}\big|x_O(y_P - y_Q) + x_P(y_Q - y_O) \\ &\quad + x_Q(y_O - y_P)\big| \\ &= \frac{1}{2}\big|(\frac{-1+\sqrt{13}}{2})(\frac{2+\sqrt{13}}{3}) \\ &\quad - (\frac{-1-\sqrt{13}}{2})(\frac{2-\sqrt{13}}{3})\big| \\ &= \frac{\sqrt{13}}{6}. \end{aligned}$$

(d) The area of the triangle OPQ is $A = \frac{h}{2} D_{PQ}$, where h is the shortest distance from O to PQ produced. So,

$$h = \frac{2A}{D_{PQ}} = \frac{2 \times \sqrt{13}/6}{13/3} = \frac{\sqrt{13}}{13}.$$

---

16. $ABCD$ is a cyclic quadrilateral and E is the point outside the circle where BA produced meets CD produced. If $CD = 5$, $DE = 8$ and $EB = 11$ determine the length AE. (Note: A **cyclic polygon** is one whose vertices lie on the circumference of a circle.)

- -

Solution:
Imagine a point P on the circumference so that EP is tangent to the circle.

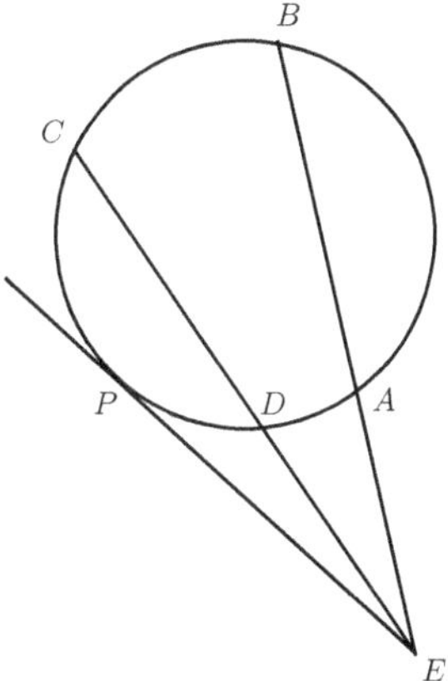

Figure 4.7: Exercise 7.

By the Tangent-Secant theorem, we have $EP^2 = EA \times EB = EC \times ED$. Thus, $EA = \dfrac{EC \times ED}{EB} = \dfrac{(8+5) \times 8}{11} = \dfrac{104}{11}$.

---

17. $ABCD$ is a cyclic quadrilateral and E is the point outside the circle where BA produced meets CD produced. If $CD = 5$, $DE = 8$, $EB = 11$ and $\angle CEB = 20°$, determine the lengths BC and AD.

- -

Solutions:
Consider the figure below,

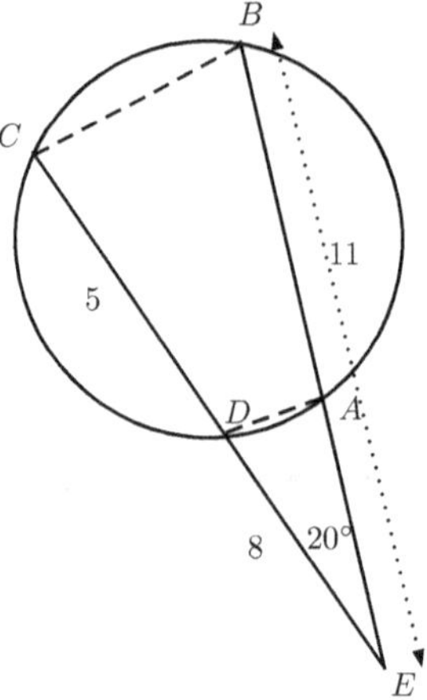

Figure 4.8: Problem 3.

BC follows from the cosine rule:

$$\begin{aligned} BC^2 &= 13^2 + 11^2 - 2(13)(11)\cos(20°) \\ BC &= 4.61\,. \end{aligned}$$

We next determine EA using the Tangent-Secant Theorem, which implies $EA \times 11 = ED \times EC$. Thus $EA = 104/11$.

Finally, AD is determined using the cosine rule,

$$\begin{aligned} AD^2 &= 8^2 + (104/11)^2 - 2(8)(104/11)\cos(20°) \\ AD &= 3.35. \end{aligned}$$

---

18. The points A, B, C and D lie cyclically on the circumference of a circle as shown below. The lines AC and BD intersect at E with $AE = 7$, $EC = 2$, $BE = 6$ and $\angle ABE = 35°$. Find

    (a) The lengths of ED and DC.

    (b) $\angle CED$.

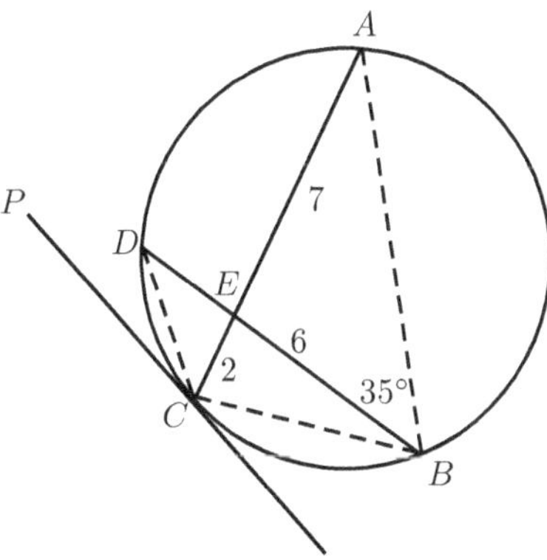

Figure 4.9: Problem 7.

- -

Solutions:

(a) From the Intersecting Chords Theorem, $6 \cdot ED = 7 \cdot 2 \Rightarrow ED = 7/3$.
Next, we have $\angle ABD = \angle ACD = 35°$ as they are subtended by the same chord (see the Inscribed Angle Theorem). Then, by using the cosine rule and solving the quadratic equation in DC, we have

$$\begin{aligned} ED^2 &= DC^2 + EC^2 - 2(DC)(EC)\cos 35° \\ \frac{49}{9} &= DC^2 + 4 - (4\cos 35°)DC \\ 0 &= DC^2 - (4\cos 35°)DC - \frac{13}{9} \\ \Rightarrow DC &= 3.67\ . \end{aligned}$$

(b) From $\triangle CDE$,

$$\begin{aligned} \frac{\sin \angle CED}{CD} &= \frac{\sin \angle ECD}{ED} \\ \angle CED &= \sin^{-1}\left(\frac{3.67 \sin 35°}{7/3}\right) \\ &= 64.44° \text{ or } 115.56°. \end{aligned}$$

As $\angle CED$ is opposite the longest side, it must be the largest angle, so we have to choose $\angle CED = 115.56°$. Hence $\angle AED = 180° - 115.56° = 64.44°$.

---

19. The points A, B, C and D lie cyclically on the circumference of a circle. The lines AB and AC are of the same length while $BC = 4$. If $\angle BDC = 20°$, find the area of triangle ABC.

Solution:
Consider the figure below,

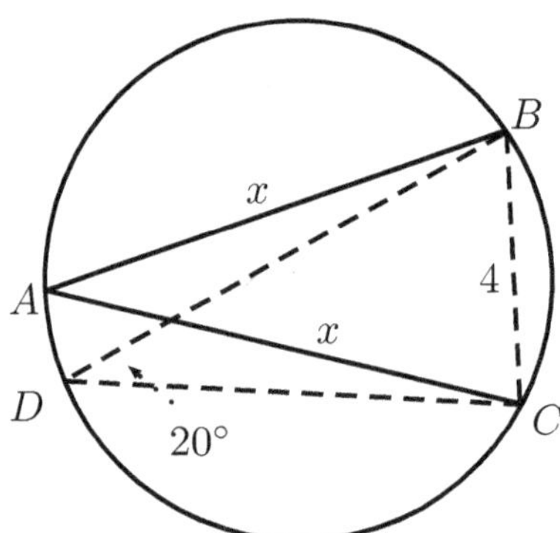

Figure 4.10: Problem 9.

$\angle BAC = \angle BDC = 20°$ as they are subtended by the same chord. Using the cosine rule in $\triangle BAC$,

$$\begin{aligned} BC^2 &= AB^2 + AC^2 \\ &\quad -2(AB)(AC)\cos\angle BAC \\ 4^2 &= 2x^2(1-\cos 20°) \\ \Rightarrow x &= \frac{4}{\sqrt{2(1-\cos 20°)}}. \end{aligned}$$

Thus the area of triangle ABC is

$$\begin{aligned} \text{Area} &= \frac{1}{2}x^2 \sin\angle BAC \\ &= \frac{1}{2} \times \left(\frac{16}{2(1-\cos 20°)}\right) \sin 20° \\ &= 22.69. \end{aligned}$$

4.5 Test Yourself

1. Which is larger, $\sec(15°)$ or $\csc(15°)$? Justify your answer without using a calculator.

2. Given $\sin x = 1/5$, and $0 < x < \pi/2$, express exactly the value of $\tan(x + \pi/6)$.

3. Find the smallest positive x which solves $\tan^2 x = 2 - 3\sec x$.

4. If A is in the first quadrant and B in the second, with $\sin A = 0.2$ and $\cos B = -0.3$, find the value of $\tan(A+B)$ correct to two decimal places without using explicit values of A or B.

5. A circle has equation $x^2 + y^2 + 2x + 4y - 4 = 0$. A square $ABCD$ is inscribed in the circle with $A = (-1, 1)$. Determine

   (a) The coordinates of the other vertices of the square.

   (b) The area of the square.

6. In Worked Example (18), determine

   (a) $\angle ADE$.

   (b) The angle between the tangent at C and the line EC.

7. Challenge: Determine the shortest distance between the circles C and C_2 in part (c) of Worked Example (13).

Did You Know?

Euler's Identity:

$$e^{i\theta} = \cos\theta + i\sin\theta\ ,$$

where $i = \sqrt{-1}$.

4.6 Answers to Test

1. $\sec 15° < \csc 15°$.

2. $\dfrac{\sqrt{3}+2\sqrt{6}}{6\sqrt{2}-1}$.

3. $x = 1.84$.

4. *Hint: Find* $\tan A$ *and* $\tan B$ *first.*
   Answer: -1.80.

5. *Hint: A sketch will clarify the situation.*
   Answer: $B = (2, -2), C = (-1, -5), D = (-4, -2)$. The area of the square is 18.

6. (a) $\angle ADE \approx 96.2°$.
   (b) $126°$.

7. *Hint: Connect the centres of the circles by a straight line and let* A *and* B *be the points on the two circles where the line intersects. Show that* AB *is the shortest distance and calculate its value.*
   Answer: $(\sqrt{6^2 + 14^2} - 6)$.

Did You Know?

Given a triangle with perimeter P and area A, we have the inequality $P^2 \geq 12\sqrt{3}A$, with equality holding for equilateral triangles.

Chapter 5

Calculus

5.1 Review: Differential Calculus

Differential calculus deals with problems involving some "rate of change". Such problems are common in science as most of the basic equations describing Nature are **differential equations**, that is, equations which contain some derivatives.

Given a function $A(t)$, one may **differentiate** it to obtain $\frac{dA}{dt}$, also referred to as the **derivative of A with respect to t**. If one plots $A(t)$ against t, then $\frac{dA}{dt}$ is the slope of the tangent to the curve.

For example, Newton's second law of motion, $F = ma$, gives the acceleration, a, of a particle if the forces acting on it are known. If $v(t)$ is the time-varying velocity of the particle, then $a \equiv dv/dt$.

5.1.1 Relations and Notation

We list here, some common formulae of differential calculus.

Let f and g be functions of x, while A, B and n denote constants in the following.

$$\frac{d}{dx}Ax^n = Anx^{n-1}\,. \tag{5.1}$$

$$\frac{d}{dx}e^x = e^x\,. \tag{5.2}$$

$$\frac{d}{dx}\ln x = \frac{1}{x}\,. \tag{5.3}$$

$$\frac{d}{dx}\sin x = \cos x\,. \tag{5.4}$$

$$\frac{d}{dx}\cos x = -\sin x\,. \tag{5.5}$$

$$\frac{d}{dx}\tan x = \sec^2 x\,. \tag{5.6}$$

$$\frac{d}{dx}(f+g) = \frac{df}{dx} + \frac{dg}{dx}. \quad (5.7)$$

$$\frac{d}{dx}(fg) = f\frac{dg}{dx} + g\frac{df}{dx}. \quad \textbf{Product Rule} \quad (5.8)$$

$$\frac{d}{dx}\frac{f}{g} = \frac{gf' - fg'}{g^2}. \quad \textbf{Quotient Rule} \quad (5.9)$$

$$\frac{d}{dx}f(g(x)) = \frac{df}{dg} \times \frac{dg}{dx}. \quad \textbf{Chain Rule} \quad (5.10)$$

$$\frac{df}{dx} = \left(\frac{dx}{df}\right)^{-1}. \quad (5.11)$$

The remarkable relation (5.2) shows the special properties of the constant $e = 2.718....$ In fact Ae^x is the only function which is its own derivative.

Notation: Second derivatives, which express the rate of change of the rate of change, are written $\frac{d}{dx}\left(\frac{dy}{dx}\right) \equiv \frac{d^2y}{dx^2}$. A convenient short-hand for $\frac{dy}{dx}$ is y', with y'' expressing the second derivative. However if the independent variable is time, t, then the "overdot" notation is usual, that is $\frac{dx}{dt}$ would be written $\dot{x}$ and the second derivative would be $\ddot{x}$.

Caution: Verbally, $\frac{dy}{dx}$ is usually spoken as "dy by dx" and for convenience we will sometimes type it as dy/dx in this book; but remember that $\frac{dy}{dx}$ is NOT "dy" divided by "dx". It is simply a notation to express the derivative of y with respect to x, nothing more nor less.

5.2 Review: Integral Calculus

The **Fundamental Theorem of Calculus** links integral calculus to differential calculus:

$$\int_a^b f(x)dx = F(b) - F(a), \quad (5.12)$$

where $F(x)$ is a function that satisfies

$$\frac{dF(x)}{dx} = f(x). \quad (5.13)$$

We see from (5.12-5.13) that the process of "integration" is essentially the reverse of differentiation. If we omit the limits of the integration region, we have an **indefinite integral** given by

$$\int f(x)dx = F(x) + C, \quad (5.14)$$

where (5.13) still holds and C is a constant of integration which can be fixed once we have more information about the problem.

Since, as mentioned earlier, many equations in science involve rates of change (derivatives), the process of integration is often needed to obtain an expression for the quantity that is changing. For example, while Newton's Second Law gives us the acceleration of an object, to get its velocity and displacement requires us to perform some integration; this is illustrated in one of the worked examples.

5.2.1 Relations and Properties

We list here some common formulae of integral calculus. Let f and g be functions of x, while a, b and n denote constants. In the formulae for indefinite integrals below, C is a constant of integration.

$$\int (a+bx)^n \, dx = \frac{(a+bx)^{n+1}}{b(n+1)} + C, \quad n \neq -1 \,. \tag{5.15}$$

$$\int \frac{1}{a+bx} \, dx = \frac{1}{b} \ln(a+bx) + C \,. \tag{5.16}$$

$$\int \sin(a+bx) \, dx = \frac{-1}{b} \cos(a+bx) + C \,. \tag{5.17}$$

$$\int \cos(a+bx) \, dx = \frac{1}{b} \sin(a+bx) + C \,. \tag{5.18}$$

$$\int e^{(a+bx)} \, dx = \frac{1}{b} e^{(a+bx)} + C \,. \tag{5.19}$$

$$\int (f+g) \, dx = \int f \, dx + \int g \, dx \,. \tag{5.20}$$

It is useful to note the following identity

$$\int_a^c f(x)dx = \int_a^b f(x)dx + \int_b^c f(x)dx \,. \tag{5.21}$$

The area under a curve $y = f(x) \geq 0$, bounded by the x-axis and the lines $x = a$ and $x = b$, is given by

$$A = \int_a^b f(x)dx \,. \tag{5.22}$$

Note that if $f(x) < 0$ within a region $x_1 \leq x \leq x_2$, the integral in that region would give a negative value and the area is then the negative of the integral.

One may also evaluate the area between a curve and the y-axis. In this case the integral would be $\displaystyle\int_{y_1}^{y_2} x \, dy$.

5.3 Worked Examples

1. Differentiate each of the following expressions with respect to x:

   (a) $3xe^{2x}$.

   (b) $\dfrac{2+3x}{x+5}$.

- -

Solutions:

(a) Using the product rule with $f(x) = x$ and $g(x) = e^{2x}$,

$$\begin{aligned}\frac{d}{dx}[3xe^{2x}] &= 3(f\frac{dg}{dx} + g\frac{df}{dx}) \\ &= 3\Big(x(2e^{2x}) + e^{2x}(1)\Big) \\ &= 3e^{2x}(2x+1).\end{aligned}$$

(b) Using the quotient rule with $f(x) = 2 + 3x$ and $g(x) = x + 5$,

$$\begin{aligned}\frac{d}{dx}\left[\frac{2+3x}{x+5}\right] &= \frac{gf' - fg'}{g^2} \\ &= \frac{(x+5)(3) - (2+3x)(1)}{(x+5)^2} \\ &= \frac{3x + 15 - 2 - 3x}{(x+5)^2} \\ &= \frac{13}{(x+5)^2}.\end{aligned}$$

---

2. Suppose that the variable x is a function of time, and y is another variable defined by $y = f(x)$ where $f(x)$ is given by the expressions of the previous exercise. Calculate, for each case of the previous exercise, $\frac{dy}{dt}$ at $x = 0$ if $\frac{dx}{dt} = 2$.

- -

Solutions:
The Chain Rule gives

$$\begin{aligned}\frac{dy}{dt} &= \frac{dy}{dx} \times \frac{dx}{dt} \\ \Rightarrow \left.\frac{dy}{dt}\right|_{x=0} &= 2 \times \left.\frac{dy}{dx}\right|_{x=0}.\end{aligned}$$

Using the results of the previous question,

(a)

$$\begin{aligned}\left.\frac{d}{dt}(3xe^{2x})\right|_{x=0} &= 2 \times \Big(3e^{2x}(2x+1)\Big)\Big|_{x=0} \\ &= 2 \times 3 \\ &= 6.\end{aligned}$$

(b)

$$\begin{aligned}\left.\frac{d}{dt}\left(\frac{2+3x}{x+5}\right)\right|_{x=0} &= 2 \times \left.\left(\frac{13}{(x+5)^2}\right)\right|_{x=0} \\ &= 2 \times \frac{13}{25} \\ &= \frac{26}{25}.\end{aligned}$$

---

3. Determine the stationary points, if any, of the following curves and their nature (maximum, minimum or inflexion). Are the extrema local or global?

   (a) $y = 2x + \dfrac{1}{1+x}$.

   (b) $y = \sin x^2 + x^2 \quad \text{for} \quad x^2 \le \pi$.

\- -

Solutions:

(a) The derivatives are

$$\begin{aligned} \frac{dy}{dx} &= \frac{d}{dx}\left[2x + \frac{1}{1+x}\right] \\ &= 2 + (-1)(1+x)^{-2} = 2 - \frac{1}{(1+x)^2}; \\ \frac{d^2y}{dx^2} &= \frac{d}{dx}\left[2 - \frac{1}{(1+x)^2}\right] \\ &= -(-2)(1+x)^{-3} = \frac{2}{(1+x)^3}. \end{aligned}$$

For stationary points, we set $\dfrac{dy}{dx} = 0$. So

$$\begin{aligned} 2 - \frac{1}{(1+x)^2} &= 0 \\ \frac{1}{(1+x)^2} &= 2 \\ (1+x)^2 &= \frac{1}{2} \\ 1+x &= \pm\frac{1}{\sqrt{2}} \\ \therefore x &= \pm\frac{1}{\sqrt{2}} - 1\,. \end{aligned}$$

Let $x_1 = \dfrac{1}{\sqrt{2}} - 1$ and $x_2 = -\dfrac{1}{\sqrt{2}} - 1$. Since $\dfrac{d^2y}{dx^2} = \dfrac{2}{(1+x)^3}$ is positive for $x > -1$ and negative for $x < -1$,

$$\begin{aligned} \left.\frac{d^2y}{dx^2}\right|_{x=x_1} &> 0 \Rightarrow \text{There is a minimum at } x_1; \\ \left.\frac{d^2y}{dx^2}\right|_{x=x_2} &< 0 \Rightarrow \text{There is a maximum at } x_2. \end{aligned}$$

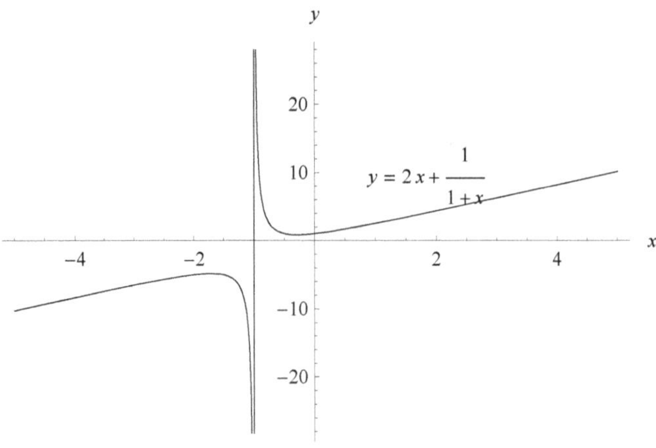

Figure 5.1: Plot of $y(x) = 2x + \frac{1}{1+x}$.

From a sketch of the graph, we see that the extrema are respectively a local minimum and a local maximum. (Note: The slope diverges at $x = -1$, which is a singular point.)

(b) The derivatives are

$$\begin{aligned}
\frac{dy}{dx} &= \frac{d}{dx}\left[\sin x^2 + x^2\right] \\
&= 2x\cos x^2 + 2x \\
&= 2x(\cos x^2 + 1); \\
\frac{d^2y}{dx^2} &= \frac{d}{dx}\left[2x(\cos x^2 + 1)\right] \\
&= 2(\cos x^2 + 1) + 2x(-\sin x^2)(2x) \\
&= 2\cos x^2 - 4x^2 \sin x^2 + 2.
\end{aligned}$$

$\frac{dy}{dx} = 0$ leads to $2x(\cos x^2 + 1) = 0 \Rightarrow x = 0$, or $\cos x^2 = -1$. Since the domain is $x^2 \leq \pi$, we have $x = 0$ or $\sqrt{\pi}$. At stationary points, the second derivatives are

$$\begin{aligned}
\left.\frac{d^2y}{dx^2}\right|_{x=0} &= 4 > 0 \\
&\Rightarrow \text{ a minimum at } x = 0. \\
\left.\frac{d^2y}{dx^2}\right|_{x=\sqrt{\pi}} &= 0 \Rightarrow \text{inconclusive.}
\end{aligned}$$

The second derivative test is inconclusive for $x = \sqrt{\pi}$. To determine the nature of this stationary point, we have to examine the first derivative in its vicinity.

x	$x=\left(\sqrt{\pi}\right)^-$	$x=\sqrt{\pi}$	$x=\left(\sqrt{\pi}\right)^+$
$\frac{dy}{dx}$	$+$	0	$+$

Figure 5.2: First derivative test at $x = \sqrt{\pi}$.

From the table, we see that the slope dy/dx before and after the point $x = \sqrt{\pi}$ has the same sign. Hence we conclude that there is a point of inflexion at $x = \sqrt{\pi}$.

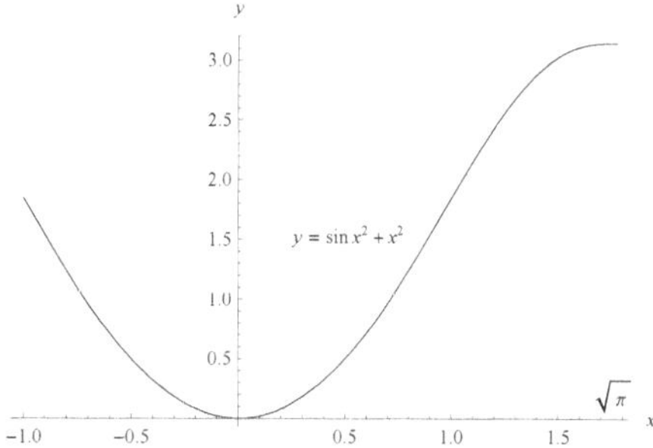

Figure 5.3: Plot of $y(x) = \sin x^2 + x^2$ for $x \leq \sqrt{\pi}$.

From the graph, we see that the stationary point at $x = 0$ is a global minimum.

---

4. Find the equation of the tangent and normal to the curve $y = x^2 - 3x + 5$ at the point $x = 1$.

- -

Solution:
Recall: If the tangent is $y = mx + c_1$, then the normal is $y = -\dfrac{1}{m}x + c_2$. Here m is the slope of the tangent and c_1, c_2 are constants.

The slope of the curve at the point $x = 1$ is

$$\frac{dy}{dx}\Big|_{x=1} = (2x-3)\Big|_{x=1} = 2(1) - 3 = -1.$$

So the tangent is $y = -x + c_1$. Now, at $x = 1$ on the curve, we have $y = 3$. Therefore the point $(1, 3)$ must be on the tangent and normal lines at $x = 1$, and we can use it to determine the constants c_1 and c_2. Substitution gives $c_1 = 3 + 1 = 4$, which fixes the equation of the tangent at that point to be $y = -x + 4$. Similarly, for the normal $y = -\dfrac{1}{m}x + c_2 = x + c_2$, we find $c_2 = 3 - 1 = 2$. Thus the equation of normal at that point is $y = x + 2$.

---

5. If the radius of an inflating sphere changes from 10 cm to 10.1 cm, use calculus to estimate the corresponding change in the quantities in parts (a-c) below.

(a) Circumference of a great circle.

(b) The surface area of the sphere.

(c) The volume of the sphere.

(d) Check the accuracy of your answers above by calculating the relevant quantities exactly.

- -

Solutions:
Recall: The approximate change in a function $y(x)$ is $\Delta y \approx \left(\frac{dy}{dx}\right) \Delta x$. Given that the radius of sphere changes from $R = 10$ cm to 10.1 cm, $\Delta R = 0.1$ cm.

(a) Since $C(R) = 2\pi R$ and $\dfrac{dC(R)}{dR} = 2\pi$,

$$\begin{aligned}\Delta C &\approx \frac{dC(R)}{dR} \times \Delta R \\ &= 2\pi \Delta R \\ &= 0.2\pi \text{ cm} \\ &= 0.628 \text{ cm}.\end{aligned}$$

(b) The surface area is $A(R) = 4\pi R^2$ and $\dfrac{dA(R)}{dR} = 8\pi R$,

$$\begin{aligned}\Delta A &\approx \frac{dA(R)}{dR} \times \Delta R \\ &= 8\pi R \Delta R \\ &= 8\pi(10 \text{ cm}) \times 0.1 \text{ cm} \\ &= 25.1 \text{ cm}^2.\end{aligned}$$

(c) The volume of the sphere is $V(R) = \dfrac{4}{3}\pi R^3$ and $\dfrac{dV(R)}{dR} = 4\pi R^2$,

$$\begin{aligned}\Delta V &\approx \frac{dV(R)}{dR} \times \Delta R \\ &= 4\pi R^2 \Delta R \\ &= 40\pi \text{ cm}^3 \\ &= 125.6 \text{ cm}^3.\end{aligned}$$

(d) We can use exact formulae to compute the changes of the quantities as follows:

(i)

$$\begin{aligned}\Delta C &= C_{R=10.1\text{cm}} - C_{R=10\text{cm}} \\ &= 2\pi\,(10.1 \text{ cm} - 10 \text{ cm}) \\ &= 0.2\pi \text{ cm}.\end{aligned}$$

In this case the circumference computed using the approximation method in part (a) is the same as that computed by the exact method. We have perfect accuracy. (Note: The perfect accuracy in this case is due to C being linear in R).
(ii)

$$\begin{aligned}\Delta A &= A_{R=10.1\text{cm}} - A_{R=10\text{cm}}\\ &= 4\pi\left(10.1^2\ \text{cm}^2 - 10^2\ \text{cm}^2\right)\\ &= 8.04\pi\ \text{cm}^2.\end{aligned}$$

The error in using the approximation method is $\left|\frac{8.04\pi - 8\pi}{8.04\pi} \times 100\%\right| = 0.50\%$. We have high accuracy.
(iii)

$$\begin{aligned}\Delta V &= V_{R=10.1\text{cm}} - V_{R=10\text{cm}}\\ &= \frac{4}{3}\pi\left(10.1^3\ \text{cm}^3 - 10^3\ \text{cm}^3\right)\\ &= 40.4\pi\ \text{cm}^3.\end{aligned}$$

The error in using the approximation method is $\left|\frac{40.4\pi - 40\pi}{40.4\pi} \times 100\%\right| = 1.0\%$. We have high accuracy.

---

6. Determine the value of the second derivative of $\sin(\ln x)$ at $x = 1$.

- -

Solution:
Use the Chain Rule and formulae for derivatives of the ln and sin functions:

$$\begin{aligned}\frac{dy}{dx} &= \frac{d}{dx}\sin\ln x\\ &= \frac{\cos\ln x}{x}\\ \left.\frac{d^2y}{dx^2}\right|_{x=1} &= \left.\left(-\frac{\sin\ln x}{x^2} - \frac{\cos\ln x}{x^2}\right)\right|_{x=1}\\ &= -1\ .\end{aligned}$$

---

7. A farmer wants to fence off a rectangular section of his farm but the material he has can only cover a boundary of length L. Determine the dimensions of the rectangular section (in terms of L) which would

   (a) Maximise the enclosed area.

(b) Minimise the enclosed area.

- -

Solutions:

Denote the length and the width of the rectangular section by l and w. The area is $A = lw$, while the perimeter gives the constraint $L = 2(l + w)$. Using the constraint to eliminate w from the first equation gives $A = \frac{Ll}{2} - l^2$ with L a constant.

(a) To maximize the enclosed area, we have to ensure $dA/dl = 0$ and $d^2A/dl^2 < 0$.

$$\frac{dA}{dl} = \frac{L}{2} - 2l\ .$$

$$\frac{dA}{dl} = 0 \Rightarrow l = \frac{L}{4}\ .$$

Furthermore, $\frac{d^2A}{dl^2} = -2 < 0$, so the solution $l = L/4$, with corresponding $w = L/4$, leads to a maximum area (a sketch of the curve shows that it is indeed a global maximum). The rectangle with maximum area is the square $L/4 \times L/4$.

We can also obtain the result by completing the square:

$$A = -l^2 + \frac{Ll}{2} = -\left(l - L/4\right)^2 + \frac{L^2}{16} \le \frac{L^2}{16}.$$

The area is maximum when the term in bracket vanishes, that is, $(l - L/4)^2 = 0 \Rightarrow l = L/4$. This method directly confirms the maximum to be global.

(b) From part (a), we see that in the limit $l \to 0$, the area $A = -(l - L/4)^2 + L^2/16$ decreases, approaching zero. Thus the area is minimised when we have a finite width with infinitesimal length, or vice-versa.

---

8. (a) Evaluate $\int_0^{2\pi} \sin x\ dx$.

(b) Find the area between the x-axis and $y = \sin x$ for $0 \le x \le 2\pi$.

(c) Why is the answer to part (b) different from that in part (a)?

- -

Solutions:

(a)

$$\begin{aligned} I &= \int_0^{2\pi} \sin x\ dx \\ &= [-\cos x]_0^{2\pi} \\ &= -\cos(2\pi) - [-\cos 0] \\ &= -1 - (-1) \\ &= 0\ . \end{aligned}$$

(b) To find the area, we first plot the curve $y = \sin x$ for $x \in [0, 2\pi]$.

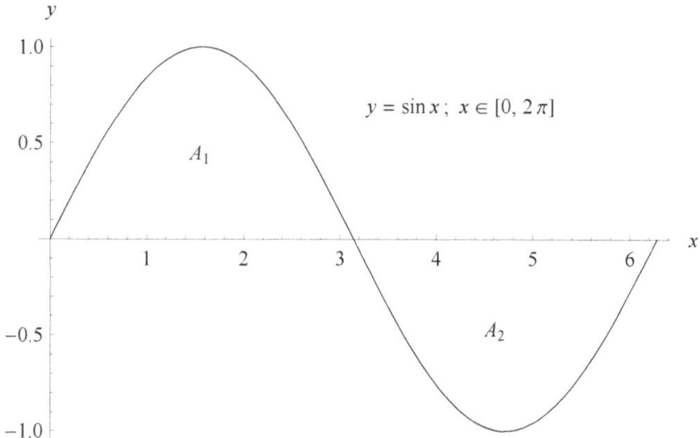

Figure 5.4: Plot of $y(x) = \sin x$.

From the graph, we see that there is a segment A_2 located below the x-axis, and the integration along that stretch would give a negative value. Therefore to get the area between the x-axis and the curve, we need to divide the integration into two parts and add the magnitudes. We have

$$\begin{aligned} A_1 &= \int_0^{\pi} \sin x \; dx \\ &= [-\cos x]_0^{\pi} \\ &= [-\cos \pi] - [-\cos 0] \\ &= 1 - (-1) \\ &= 2. \end{aligned}$$

$$\begin{aligned} A_2 &= \left| \int_{\pi}^{2\pi} \sin x \; dx \right| \\ &= \left| [-\cos x]_{\pi}^{2\pi} \right| \\ &= \left| [-\cos 2\pi] - [-\cos \pi] \right| \\ &= \left| -1 - (1) \right| \\ &= 2. \end{aligned}$$

Therefore $A = A_1 + A_2 = 2 + |-2| = 4$.

(c) It is clear that in part (a) half of the integral contributes a negative amount which exactly cancels the positive contribution from the other half.

---

9. Integrate each of the following with respect to x:

(a) $3x^2(1+x) - \dfrac{2}{x}$.

(b) $7\sin 2x \cos 2x$.

- -

Solutions:

(a) Using the standard formulae from Sect.(1.1.1),

$$\begin{aligned}\int \left(3x^2(1+x) - \frac{2}{x}\right)\, dx &= \int \left(3x^2 + 3x^3 - \frac{2}{x}\right)\, dx \\ &= \frac{3x^{2+1}}{(2+1)} + \frac{3x^{3+1}}{(3+1)} - 2\ln x + C \\ &= x^3 + \frac{3x^4}{4} - 2\ln x + C.\end{aligned}$$

(b) Use $2\sin x \cos x = \sin 2x$,

$$\begin{aligned}\int 7\sin 2x \cos 2x\; dx &= \frac{7}{2}\int \sin 4x\; dx \\ &= \frac{7}{2}\left[-\frac{\cos 4x}{4}\right] + C \\ &= -\frac{7\cos 4x}{8} + C.\end{aligned}$$

---

10. Evaluate each of the following definite integrals to two decimal places:

(a) $\displaystyle\int_1^2 dx \left(3x^2(1+x) + \frac{2}{x}\right)$.

(b) $\displaystyle\int_0^{\pi} \sin 2x \cos 2x\; dx$.

- -

Solutions:

(a)

$$\begin{aligned}\int_1^2 \left(3x^2(1+x) + \frac{2}{x}\right)\, dx &= \left[x^3 + \frac{3x^4}{4} + 2\ln x\right]_1^2 \\ &= \left[2^3 + \frac{3(2^4)}{4} + 2\ln 2\right] - \left[1 + \frac{3}{4} + 2\ln 1\right] \\ &= 7 + \frac{45}{4} + 2\ln 2 \\ &\approx 19.64\ .\end{aligned}$$

(b)

$$\begin{aligned}\int_0^\pi \sin 2x \cos 2x \; dx &= \frac{1}{7}\left[-\frac{7\cos 4x}{8}\right]_0^\pi \\ &= \left[-\frac{\cos 4\pi}{8}\right] - \left[-\frac{\cos 4(0)}{8}\right] \\ &= -\frac{1}{8} - (-\frac{1}{8}) \\ &= 0 \; .\end{aligned}$$

(Note: Compare this result with Example 8).

---

11. If $\displaystyle\int_1^6 f(x)dx = -20$ and $\int_1^8 f(x)dx = 10$,

(a) Find $\displaystyle\int_6^8 f(x)dx$.

(b) What can you say about the sign of $f(x)$ for $6 \leq x \leq 8$?

- -

Solutions:

(a) Using Eq.(5.21),

$$\begin{aligned}\int_1^8 f(x)\; dx &= \int_1^6 f(x)\; dx + \int_6^8 f(x)\; dx \\ \therefore \int_6^8 f(x)\; dx &= \int_1^8 f(x)\; dx - \int_1^6 f(x)\; dx \\ &= 10 - (-20) \\ &= 30 \; .\end{aligned}$$

(b) Since $\int_6^8 f(x)\; dx = 30 > 0$, the function $f(x)$ has to be positive for parts of the interval $[6, 8]$ (not necessarily over the whole interval).

---

12. Find the area between $y = \dfrac{1}{\sqrt{x}}$ and the x-axis over the range $0 \leq x \leq 1$.

- -

Solution:
Note that $\dfrac{1}{\sqrt{x}} = x^{-1/2}$. Then, using the standard formula,

$$\int_0^1 \frac{1}{\sqrt{x}}\, dx = \Big[2\sqrt{x}\Big]_0^1 = 2\,.$$

---

13. Integrate $\dfrac{5x^2+2x+1}{x+1}$ with respect to x.

- -

Solution:
Performing a long division and a partial fraction decomposition of the integrand, we have (see Volume 3 of this series)

$$\begin{aligned} \int \frac{5x^2+2x+1}{x+1}\, dx &= \int (5x-3)\, dx + \int \frac{4}{(x+1)}\, dx \\ &= \frac{5}{2}x^2 - 3x + 4\ln(x+1) + C. \end{aligned}$$

---

14. Sketch the curves $y = 2 - 3x^2$ and $y = x^2$ and determine the area bounded by them.

- -

Solution:
The curves are sketched below:

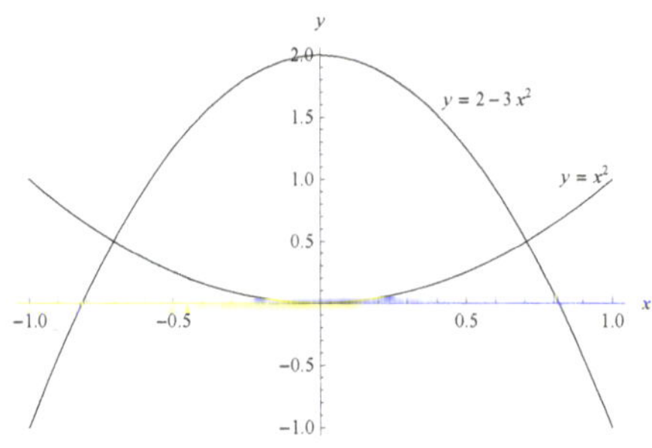

Figure 5.5: Plot of $y = 2 - 3x^2$ and $y = x^2$.

The intersection of the two curves is at $x = \pm\frac{1}{\sqrt{2}}$. The area bounded is thus the difference between the areas bounded by each curve and the x-axis between the lines $y = x = \pm\frac{1}{\sqrt{2}}$.

$$\begin{aligned} A &= \int_{-\frac{1}{\sqrt{2}}}^{\frac{1}{\sqrt{2}}} (2 - 3x^2)\ dx - \int_{-\frac{1}{\sqrt{2}}}^{\frac{1}{\sqrt{2}}} x^2\ dx \\ &= \int_{-\frac{1}{\sqrt{2}}}^{\frac{1}{\sqrt{2}}} (2 - 4x^2)\ dx \\ &= \left[2x - \frac{4}{3}x^3\right]_{-\frac{1}{\sqrt{2}}}^{\frac{1}{\sqrt{2}}} \\ &= \frac{8}{3\sqrt{2}} \\ &= \frac{4\sqrt{2}}{3}. \end{aligned}$$

---

15. A particle moves along the x-axis with velocity $v(t) = 2 - t^2 + 3t$.

    (a) If $x = 2$ when $t = 0$, determine $x(t)$.

    (b) At what time $t_1 > 0$ does the particle re-appear at $x = 2$?

    (c) At what time t_2 is the particles' velocity instantaneously zero?

    (d) Sketch $x(t)$ versus t for $t > 0$.

    (e) What distance has the particle travelled between times t_1 and t_2?

- -

Solutions:

(a) Since $v(t) \equiv \frac{dx}{dt}$, we integrate $v(t)$ with respect to t to obtain $x(t)$:

$$\begin{aligned} x(t) &= \int v(t)\ dt \\ &= \int (2 - t^2 + 3t)dt \\ &= 2t - \frac{t^3}{3} + \frac{3}{2}t^2 + C. \end{aligned}$$

When $t = 0$, we have $x = 2$, which fixes $C = 2$. The displacement is

$$x(t) = 2 + 2t + \frac{3}{2}t^2 - \frac{t^3}{3}.$$

(b) We have to solve $x(t) = 2 + 2t + \frac{3}{2}t^2 - \frac{t^3}{3} = 2$ for t.

This implies $-\frac{t}{6}\left(2t^2 - 9t - 12\right) = 0$, which has solutions $t = 0$, $\frac{9 \pm \sqrt{177}}{4}$.

The positive solution is $t_1 = \frac{9 + \sqrt{177}}{4} \approx 5.576$.

(c) For $v(t) = 2 - t^2 + 3t = 0$, we need

$$\begin{aligned} 2 - t^2 + 3t &= 0 \\ t &= \frac{-3 \pm \sqrt{9 - 4(-1)(2)}}{-2} \\ &= \frac{3 \mp \sqrt{17}}{2} \end{aligned}$$

Taking the positive solution, $t_2 = \frac{3 + \sqrt{17}}{2} \approx 3.562$.

(d) The plot is:

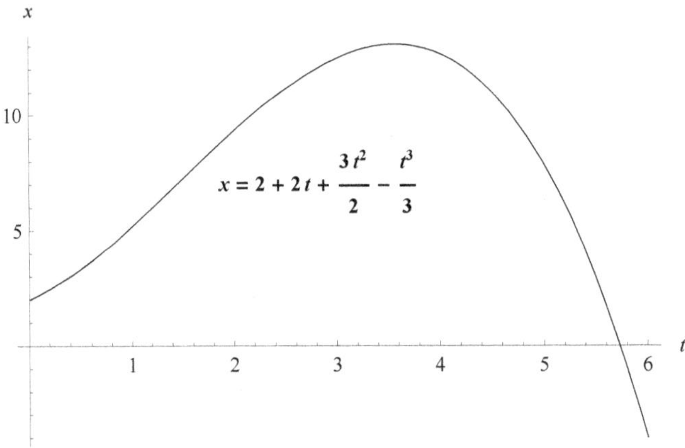

Figure 5.6: Plot of $x = 2 + 2t + \frac{3}{2}t^2 - \frac{t^3}{3}$.

(e) From part (d), we see that from t_2 to t_1 the displacement is always positive, so the distance moved is given by $D = |x(t_2) - x(t_1)| \approx 11.09$. We can also obtain the distance by integrating over the velocity vector, that is, $\left|\int_{t_1}^{t_2} (2-t^2+3t)\, dt\right| \approx 11.09$.

---

16. It is known that $y(x) = x + \ln(x + 1)$ and that x is a function of t. If $dx/dt = 2t$ and $x = 1$ at $t = 0$, calculate dy/dt at $t = 2$.

- -

Solution:
Given $\frac{dx}{dt} = 2t \Rightarrow x = t^2 + C$. As $x = 1$ when $t = 0$, we fix $C = 1$ and $x = t^2 + 1$. Then, by the Chain Rule,

$$\begin{aligned} \frac{dy}{dt} &= \frac{dy}{dx} \cdot \frac{dx}{dt} \\ &= \left(1 + \frac{1}{x+1}\right) \cdot (2t) \\ &= \left(1 + \frac{1}{t^2+2}\right) \cdot (2t) . \\ \left.\frac{dy}{dt}\right|_{t=2} &= \frac{14}{3} . \end{aligned}$$

---

17. Given a curve $y = f(x)$, its length between two points may be calculated using the formula $L = \int_a^b dx \sqrt{1 + (y')^2}$ where $y' = dy/dx$.

 Verify the formula for the straight line $y = mx$ between $0 \le x \le 1$.

- -

Solution:
For a straight line $y = mx$, we have $y'(x) = m$, so

$$\begin{aligned} L &= \int_0^1 dx \sqrt{1 + m^2} \\ &= \sqrt{1 + m^2}\, \left[x\right]_0^1 \\ &= \sqrt{1 + m^2} . \end{aligned}$$

Now, at $x = 1$ on the straight line we have $y = m$. The distance between the points $O(0,0)$ and $P(1,m)$ may be calculated using Pythagoras' theorem: $\sqrt{(1-0)^2 + (m-0)^2} = \sqrt{1 + m^2}$, which agrees with the result obtained above using the formula for L.

---

18. If $y = xe^x$, find dy/dx. Hence, or otherwise, deduce $\int_0^1 xe^x \, dx$.

- -

Solution:
Direct differentiation of $y = xe^x$ gives $dy/dx = xe^x + e^x$. Thus, re-arranging and using

Eq.(5.20),

$$\begin{aligned}
\int_0^1 xe^x \, dx &= \int_0^1 \left(\frac{dy}{dx} - e^x\right) dx \\
&= \int_0^1 \frac{dy}{dx}\, dx - \int_0^1 e^x \, dx \\
&= \left[xe^x\right]_0^1 - \left[e^x\right]_0^1 \\
&= 1.
\end{aligned}$$

5.4 Test Yourself

1. Determine the stationary points, if any, of the curve $y = (x+1)^2 e^{-x}$ and their nature (maximum, minimum or inflexion). Are the extrema local or global?

2. Find the equation of the tangent and normal to the curve $y = x^3 + x^2 + 2x + 1$ at the point $x = 1$.

3. Determine the value of the second derivative of $3x^2(1+x) + 2\ln x$ at $x = 1$.

4. If the volume of an inflating sphere changes from 50 cm^3 to 51 cm^3,

   (a) Use calculus to estimate the change in its radius.

   (b) Check the result in part (a) by computing the corresponding radii for the two spheres exactly.

5. Integrate $\left(\frac{1}{e^{2x}} - \frac{1}{\sqrt{2+3x}}\right)$ with respect to x.

6. Find the area between $\sin^2 x$ and the x-axis over the range $0 \le x \le \pi/2$.

7. Given a curve $y = f(x)$, its length between two points may be calculated using the formula $L = \int_a^b dx \sqrt{1 + (y')^2}$ where $y' = dy/dx$. Calculate the length of the curve $y = x^{3/2}$ between $0 \le x \le 1$.

8. Challenge: Sketch the three curves $y = (x-1)^2 + 1$, $y = 2$ and $y = 4$, and determine the area bounded by them.

Did You Know?

$$\int_{-\infty}^{+\infty} e^{-x^2} dx = \sqrt{\pi}$$

5.5 Answers to Test

1. Global minimum at $x = -1$. Local maximum at $x = 1$.

2. $T : y = 7x - 2$; $N : 7y + x - 36 = 0$.

3. 22.

4. $\Delta R = 0.015$ cm.

5. $-\frac{1}{2e^{2x}} - \frac{2\sqrt{2+3x}}{3} + C.$

6. *Hint: Use* $\cos 2x = 1 - 2\sin^2 x$. Answer: $\frac{\pi}{4}$.

7. ≈ 1.44.

8. *Hint: It might be easier to first determine the area bounded between* $y = (x-1)^2 + 1$ *and* $y = 4$, *and separately the area bounded between* $y = (x-1)^2 + 1$ *and* $y = 2$.

   Answer: $4\sqrt{3} - \frac{4}{3}$.

Did You Know?

The Koch snowflake is a planar figure with finite area but infinite perimeter!

How is that possible?
The boundary of the Koch Snowflake is "fractal", with no smooth point.

Fractals are usually included in the study of complex systems.

Chapter 6

Real World Applications

6.1 Worked Examples

1. When two resistors R_1 and R_2 are connected in parallel in a electrical circuit, their combined resistance R is determined by $\frac{1}{R} = \frac{1}{R_1} + \frac{1}{R_2}$. All resistances are positive.

   (a) Prove that $R \leq R_1$ and $R \leq R_2$.

   (b) Express R as a rational function of R_1 and R_2.

- -

Solutions:

(a) Since

$$\frac{1}{R} = \frac{1}{R_1} + \frac{1}{R_2}, \tag{6.1}$$

and all variables are positive, so $\frac{1}{R} \geq \frac{1}{R_1}$ and $\frac{1}{R} \geq \frac{1}{R_2}$. Therefore, $R \leq R_1$ and $R \leq R_2$.

(b) Re-arranging the terms: $\frac{1}{R} = \frac{R_2 + R_1}{R_1 R_2}$. Hence, $R = \frac{R_1 R_2}{(R_1 + R_2)}$.

---

2. You are instructed to construct an open top box from a square, $a \times a$, piece of cardboard: First cut out squares of dimensions x by x at each corner of the cardboard, then fold the cardboard to form a box of dimensions x by $(a - 2x)$ by $(a - 2x)$.

   (a) Find the values of x which make the volume of the box a minimum and explain the physical situation those values correspond to.

   (b) Hence, or otherwise, explain why for some value of x the volume of the box would be a maximum.

   (c) Determine the maximum volume using calculus.

- -

Solutions:

(a) The volume of the box is $V(x) = x(a-2x)^2$. As physically $V \geq 0$, and $(a-2x)^2 \geq 0$, so we see by inspection that the minimum volume corresponds to the situation $x = 0$ (no height) or $x = a/2$ (no base area). Both minima correspond to zero volume.

(b) As $V(x) \geq 0$ is a smooth function, it must have a maximum in the physical range $0 < x < a/2$ since it reaches its minimum value (zero) at the end points.

(c) The turning points satisfy the equation

$$\frac{dV}{dx} = a^2 - 8ax + 12x^2 = 0,$$

which gives us the values $x = a/2$ and $x = a/6$. To determine the nature of the these extrema (minima or maxima), we need to check the value of $d^2V/dx^2 = -8a + 24x$. For $x = a/2$, we have $d^2V/dx^2 > 0$, so this is a minimum point (as we already knew from part (a)). For the other extremum at $x = a/6$, we have $d^2V/dx^2 = -4a$, so $x = a/6$ is at least a local maximum. Since at the boundaries $x = 0$ and $x = a/2$, the smooth function $V(x)$ does not exceed its value at the local maximum, so the local maximum is actually a global maximum.
The maximum volume is given by $V(x = a/6) = 2a^3/27$.

---

3. The pH value of a solution indicates its hydrogen ion concentration, denoted by $[H^+]$ in moles per litre, through the formula

$$pH = -\log_{10}[H^+] \ .$$

Neutral solutions have a pH of 7, acidic solutions have pH values below 7, while base solutions have a pH above 7.

(a) Determine the hydrogen ion concentration of pure water, a neutral solution.

(b) Do acidic solutions have a larger, or smaller, hydrogen ion concentration than neutral solutions?

- -

Solutions:

(a) Neutral solution means pH=7, so $7 = -\log_{10}[H^+]$, or $[H^+] = 10^{-7}$ moles per litre.

(b) We have $[H^+] = 10^{-(\text{pH})}$. Since acidic solutions have a pH less than 7, so they will have a higher hydrogen ion concentration than a neutral solution.

---

4. The curve formed by a freely hanging cable supported only at its ends is called a catenary. Its equation is $y = \frac{a}{2}\left(e^{x/a} + e^{-x/a}\right)$. If the lowest point of the cable reaches 100 m below its support level, and if $a = 10$ m, determine the coordinates, in metres, of the support point.

- -

Solution:

For convenience, we first define the hyperbolic cosine function $\cosh(x) \equiv \left(\frac{e^x + e^{-x}}{2}\right)$. Then, we have $y = a\cosh(x/a)$. Similarly, we define $z \equiv a\sinh(x/a)$, where $\sinh(x) \equiv \left(\frac{e^x - e^{-x}}{2}\right)$ is the hyperbolic sine function. It can be easily verified that $\frac{d\cosh(x)}{dx} = \sinh(x)$. The minimum point of y is at where $\frac{dy}{dx} = 0$, which is satisfied only when $x = 0$. Therefore, we have $y = a = 10\,\text{m}$ at the minimum. The support level is 100 m above the minimum, that is, at $y = 110\,\text{m}$.

The x-coordinates of the two support points are obtained by solving $110 = 10\cosh(x/10)$. Explicitly, $x = \pm 10\cosh^{-1}(11) = \pm 30.89$, where $\cosh^{-1}()$ is the inverse hyperbolic cosine function. The $\pm$ sign comes from the fact that y is unchanged upon the substitution $x \to -x$.

Ans: $x = \pm 30.89,\ \ y = 110$.

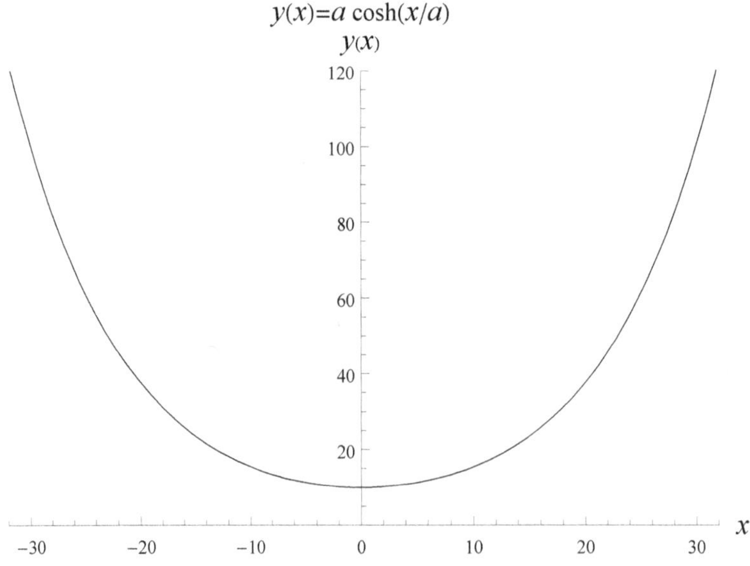

Figure 6.1: Plot of $y(x)$, with $a = 10$.

---

5. A car can reach a maximum speed of 80 km/hr. Its fuel consumption (in some units) when moving at v km/hr is $C = (v/20)^3$ units per hour. The car must be driven from

one town to another, a distance of 300 km, in less than 6 hours, but the total fuel consumption must not exceed 150 units. Assuming, for simplicity, that the car travels at constant speed, determine the range of values for v which achieve the objective.

– –

Solution:
We have $v = d/t$ where $d = 300$ is the distance travelled and t the time in hours. So $t < 6 \Rightarrow 300/v < 6$, that is $v > 300/6 = 50$ km/h.
The fuel consumption rate is $C = (v/20)^3$. The total fuel consumption after t hours is $F = C \times t = Cd/v$. We require $F < 150$, which implies $300v^2/8000 < 150$, or $v < 63.25$.

Ans: $50 < v < 63.25$.

---

6. Eona determines the height of a tree using a metre rule placed vertically on the ground: She finds that the shadow of the metre rule is 1.3m long when the shadow of her tree is 5.5m long. How tall is Eona's tree? [Ans: 4.23 m]

– –

Solution:
As shown in the figure, the angle is the same. That is, we have $\tan(\theta) = 1/1.3 = x/5.5$, so $x = 5.5/1.3 = 4.23\,\text{m}$.

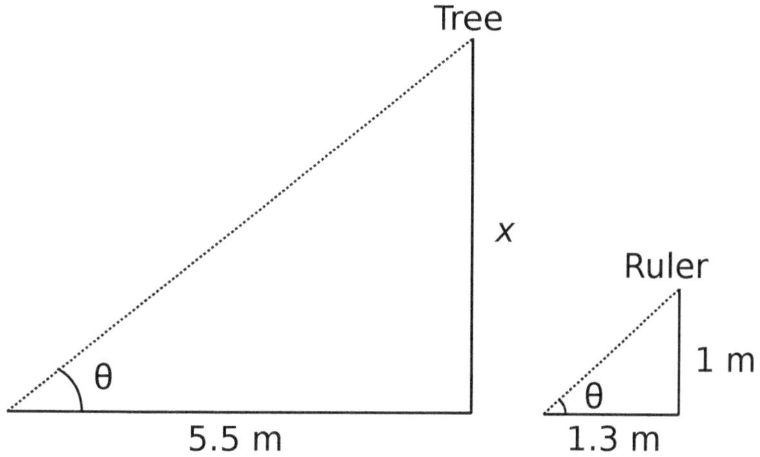

Figure 6.2: Similar Triangles.

---

7. A guitar string is tied at its two ends located at $x = 0$ and $x = L$. When plucked, it creates "standing waves" whose vertical displacement is given by $y(x) = A\sin\frac{2\pi x}{\lambda}$ where A is the maximum amplitude and λ is the wavelength of the wave.

   (a) Using the condition $y(L) = 0$ (since the tied end does not move), find the possible values for λ in terms of a positive integer n.

   (b) For $n = 2$, locate the points of the standing wave with zero amplitude (called "nodes") and the points with maximum amplitude (called "anti-nodes").

\- -

Solutions:

(a) $y(L) = A\sin\frac{2\pi L}{\lambda} = 0 \Rightarrow 2\pi L/\lambda = n\pi$, that is $\lambda = 2L/n$ with n a positive integer.

(b) For $n = 2$, $y(x) = A\sin\frac{2\pi x}{L}$. The nodes occur at values of $x \le L$ that satisfy $y = 0$, that is $2\pi x/L = m\pi$ with m a positive integer. So the nodes are at $x = 0,\ L/2,\ L$. Similarly, the anti-nodes occur when $|y| = A$, that is at $x = L/4,\ 3L/4$.

---

8. An object with initial speed v is brought to stop in a distance d by a constant opposing force. Theory predicts that $d \propto v^2$. If the opposing force is kept constant while v is varied, suggest a plot involving d and v that would produce a straight line.

\- -

Solutions:
We have $d = kv^2$, where k is a constant. For a straight line plot involving d and v, we can either

(a) Plot d against v^2, such that the plot will have a gradient of k, or

(b) Plot $\log d$ against $\log v$. In this case, $\log d = \log k + 2\log v$, so the gradient will be 2.

---

9. The von Bertalanffy model for tumour growth is given by the equation

$$\frac{dV}{dt} = aV^p - bV^q\,,$$

where V is the size of the tumour, t the time and a, b, p, q are positive constants.

(a) The two terms on the right hand side of the equation represent growth and degradation. Identify those respective terms.

(b) Show that for $q > p$ there is a maximum size to the tumour.

– –

Solutions:

(a) We are given $\dfrac{dV}{dt} = aV^p - bV^q$ where a, b, p, q are positive constants. V represents the size of the tumour, and is therefore positive. Thus, both the terms aV^p and bV^q will be non-negative. $\dfrac{dV}{dt}$ is the rate of change in the tumour size, and hence the positive term aV^p represents growth, while the negative term $-bV^q$ is for degradation.

(b) We want to show that for $q > p$, there is always a maximum size to the tumour. We first re-write $\dfrac{dV}{dt} = aV^p\left[1 - \left(\dfrac{b}{a}\right)V^{q-p}\right]$. Local extrema occur when the derivative vanishes, that is when $V = 0$ or $V \equiv V_c = \left(\dfrac{a}{b}\right)^{1/(q-p)}$.

We first note that as $V \to 0$, $dV/dt \approx aV^p$ as it will be dominated by the smaller power of V. Since $V \geq 0$, this implies that near $V = 0$, dV/dt is positive, and V increases as it deviates from zero. Thus $V = 0$ is a minimum point.

Next, we re-write the derivative again as $\dfrac{dV}{dt} = aV^p\left[1 - \left(\dfrac{V}{V_c}\right)^{q-p}\right]$. From this we see that the term in square-brackets is positive for $V < V_c$ and negative for $V > V_c$. That is, dV/dt is positive for $0 < V < V_c$ and negative for $V > V_c$.

We conclude that for $V > 0$, the tumour increases in size up to a maximum of V_c.

---

10. A thin-skinned rubber sphere is inflated by pumping it with water (which is essentially incompressible). Water is pumped in at the constant rate of 3 cm^3 s^{-1}. If the volume of the sphere is 200 cm^3 when $t = 10$, determine

(a) The radius of the sphere at $t = 1$.

(b) The rate of change of the radius at $t = 1$.

– –

Solutions:

(a) We are given $\dfrac{dV}{dt} = 3$. So $V = \int dV = \int \frac{dV}{dt}\, dt = \int 3dt = 3t + C$ where C is the integration constant. Since $V = 200$ at $t = 10$, so $C = 170$ and $V = 3t + 170$. Hence $V(t = 1) = 173$, and using $V = 4\pi r^3/3$ gives us $r = 3.457$ cm.

(b) From the expression $V = 4\pi r^3/3$, we get $dV/dr = 4\pi r^2$. We also have from the chain rule,

$$\frac{dV}{dt} = \left(\frac{dV}{dr}\right)\left(\frac{dr}{dt}\right) = 4\pi r^2\left(\frac{dr}{dt}\right)$$

Therefore,

$$\frac{dr}{dt} = \frac{3}{4\pi r^2} = 0.02 \text{ cm s}^{-1}$$

where we used $r = 3.457$ from the first part.

6.2 Test Yourself

1. In Worked example (1), if the value of R_1 is fixed while R_2 is a variable resistor, determine the range of values for R.

2. In Worked example (2), use a plot to estimate the maximum volume of the box when $a = 2$. Compare with the result using calculus.

3. In Worked example (3), if a solution has a hydrogen ion concentration of 5×10^{-7} moles/litre, determine its pH value. Is the solution acidic or alkaline (basic)?

4. In Worked example (6), what was the elevation of the Sun when Eona did her measurement?

5. In Worked example (10), what is the rate of change of the surface area at $t = 1$.

6. Challenge: Show that near its lowest point, the catenary mentioned in worked example (4) can be approximated by a parabola.

Did You Know?

There are an infinite number of integral solutions corresponding to Pythagoras' Theorem, $x^2 + y^2 = z^2$.

Explicitly, $x = m^2 - n^2$, $y = 2mn$, $z = m^2 + n^2$, where m, n are any positive integers.

6.3 Answers to Test

1. $0 \leq R \leq R_1$.

2. $V_{max} = 16/27$ at $x = 1/3$.

3. 6.3, acidic.

4. 37.6°.

5. 1.7375 cm s^{-2}.

6. The bottom of the catenary is at $x = 0$. Near its bottom, we can thus expand y as $y = \dfrac{(1 + x + x^2/2 \cdots) + (1 - x + x^2/2 + \cdots)}{2a} \approx \dfrac{2 + x^2}{2a}$, which is a parabola.

Did You Know?

A freely hanging cable takes the shape of a catenary, see worked example (4). Near its lowest point, a catenary can be approximated by a parabola (see test question (6)).

However, if the cable supports a bridge, and the weight of the cable is negligible compared to the weight it supports, then the shape of the cable is closer to a parabola than a catenary.

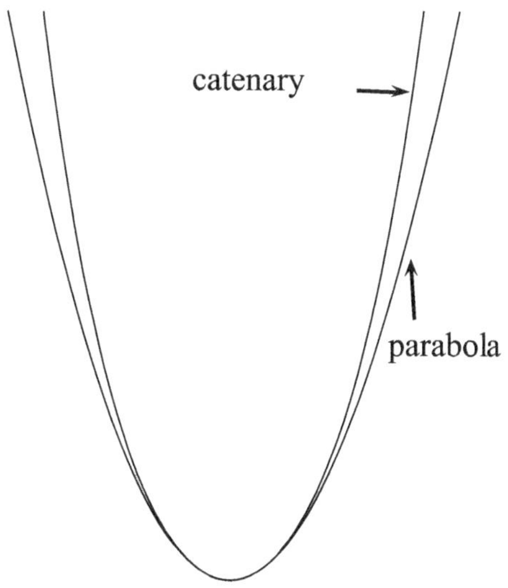

www.ingramcontent.com/pod-product-compliance
Lightning Source LLC
Chambersburg PA
CBHW081200180526
45170CB00006B/2171

* 9 7 8 1 5 4 6 9 1 0 6 4 0 *